SPRAGUE ELECTRIC

An Electronics Giant's Rise, Fall, and Life after Death

By John L. Sprague

Copyright © 2015 by John L. Sprague
All rights reserved.

ISBN: 150338781X
ISBN 13: 9781503387812

Library of Congress Control Number: 2015900299
CreateSpace Independent Publishing Platform
North Charleston, South Carolina

No part of this work may be reproduced, stored in any retrieval system, or transmitted in any form or by any means, electronic, mechanical, photocopying, microfilming, recording, or otherwise, without express written permission from the author.

CONTENTS

Introduction	v
Acknowledgements	xi
I: Sprague Specialties Company	1
II: Sprague Electric Joins the Solid-State Revolution	41
III: Progress, then Struggle to Survive, and Life after Death	101
Epilogue: John Sprague's Reflections on Robert Sprague	145

Appendices:

Appendix 1: Sprague Electric Financial Tables	161
Appendix 2: Sprague Electric Planar Process Timeline	165
Appendix 3: Robert C. Sprague and Nuclear Deterrence	167
Appendix 4: Chronology of Important Dates in Robert C. Sprague's Life	173
Endnotes	177
Index	181

INTRODUCTION

The rise of the Sprague Electric Company from a kitchen-table high-tech startup with a niche electronic product is representative of much of the U.S. electronics industry. Sprague Electric Company rose to become a thriving manufacturer employing thousands of workers, building a broad product line with international sales and a reputation for the highest quality. It then declined, went through a series of acquisitions, and eventually dissolved. Its principal manufacturing plant is now an art museum.

Sprague Electric provides a valuable business and technological history, and a cautionary tale, which serves as a lens for the stories of thousands of companies all over the world. It is the story of corporate success, and what to avoid. The Sprague Electric story portrays the value of investment in research and development, and also the effects of raw material supply chains on product lines. It is a story of a company's relations with the town where its factories were located, North Adams, Massachusetts, a small New England mill town tucked in the very northwest corner of the state, and how labor relations — initially cordial— later soured. It is a story of how a vulnerable company weathered the stresses of the Great Depression and triumphed over them, only to be brought down by the recessions of the 1970s and 1980s.

It is a history of acquisitions, mergers, and spin-offs— some of them botched— and of the strategic and tactical mistakes that eventually caused the company to vanish.

Nevertheless, Sprague Electric's successor companies continue its legacy in the electronic components industry. Corporations formed from its different business units and operations are now spread around the world.

Sprague Specialties (changed to Sprague Electric in April 1944) began in 1926 in the Quincy, Massachusetts kitchen of a young naval Officer, Ensign Robert C. ("R. C.") Sprague. At least part of Robert Sprague's brilliance came from his father, electrical inventor Frank J. Sprague who was the father of electric railroads and also made contributions to safety control technologies. Although Sprague Specialties' first product— a radio tone control—had failed, its core "Midget" capacitor served as the foundation of what became one of the world's most successful electronic component suppliers.

Seeking additional space and labor for its growing consumer oriented capacitor lines, Sprague Electric expanded and moved to North Adams, Massachusetts in 1929. During World War II, it fabricated such varied products as gas masks for civilian distribution and casings for incendiary bombs used during James Doolittle's April 1942 air raid. Crucial components for the VT proximity fuze that helped win the war flowed off Sprague Electric's North Adams production lines. When the Second World War ended with the horror of the nuclear destruction of Hiroshima and Nagasaki, special energy storage capacitors made in North Adams were part of the trigger mechanisms of the bombs; they continued to be used in America's growing nuclear arsenal. In recognition of Sprague Electric's accomplishments, the factories involved received an unheard of five Army-Navy "E" Awards, as well as a Navy Department Bureau of Ordinance "E" for their VT proximity fuze contributions. Five hundred and forty-three factory employees served in the armed forces; eighteen of those gave their lives.

World War II, the Korean War, the Cold War, and the Space Age solidified Sprague's position as a crucial supplier of high-performance and reliable military-grade components. In 1960, the company received a $1.3 million contract from Autonetics to develop super high-reliability components for the Minuteman ICBM guidance and control system. Sprague's new solid tantalum capacitor not only filled the bill admirably, it also filled a need in the expanding computer market. IBM eventually became Sprague's largest customer (Delco Radio, the electronics division of

General Motors, and AT&T were the next two largest). There were more than 50,000 Sprague devices in every *Apollo* mission and more than 25,000 in every Space Shuttle, not to mention thousands more in the related ground control equipment. Sprague also invented the multi-layer ceramic capacitor that came to dominate the world capacitor industry, but unfortunately Sprague was never able to make it a business success.

The 1947 invention of the transistor at Bell Telephone Labs, followed during the 1950s by the development of the integrated circuit, generated a revolution in electronics. Sprague Electric played an early pivotal role with Kurt Lehovec's invention of PN junction isolation, without which integrated circuits would not be possible. The development of the ion implantation process by two Sprague researchers made possible the early success of Mostek, at one time a minority subsidiary of Sprague Electric. Nevertheless, Sprague Electric had a long and rocky road to success in semiconductors which didn't occur until the early-1970s after the company had settled on a niche strategy of offering unique circuits for consumer applications, Hall Cells, and power integrated circuits. The last two remain key strategies today, but for a legacy company under very different ownership.

Despite these successes, Sprague Electric ceased to exist as an entity by 1992, although many of its former business units continued operations under different management and ownership.

INTRODUCTION

Sprague Electric's company town, North Adams, Massachusetts, was originally a Native American settlement. In 1754, European fur traders arrived who first bartered with, and then fought, the native landholders, and eventually each other. Crude dams and primitive mills appeared later, clinging precariously to the shores of the raging north and south branches of the Hoosic River. Textile, shoe, and other factories began to replace the earlier structures, initially using water power to drive their equipment. As different immigrant groups populated the workforce, the surrounding community evolved into a collection of ethnic villages. Beginning from scratch in the middle 1800s, the Arnold Print Works began construction of a massive complex of some twenty-six buildings, most of which still stand, where the Hoosic River branches converge at 87 Marshall Street.

By 1900, Arnold had become one of the world's leading textile printing corporations. Hard times came next, the result of an unfortunate management gamble on cotton futures and increasingly effective competition from southern based factories. Then Arnold Town, as some called North Adams during the Arnold ascendency, began its transition to Sprague Town, after Sprague Specialties moved to North Adams at the beginning of the great depression.

Sprague Electric located first on Beaver Street, then purchased a vacant mill on Brown Street, and during World War II replaced bankrupt Arnold Print as the owner and tenant of 87 Marshall Street. Over time,

Sprague became an electronic component giant. At its peak in the mid-1980s, it had revenues exceeding $500 million and employed 13,000 people in some twenty-four locations around the world. Nonetheless, difficulties had begun to surface twenty years earlier, beginning with increasing labor problems in North Adams. Several questionable strategic decisions and increasingly strong Asian competition made matters worse. Between 1976 and 1984, Sprague was acquired twice, and the new owners moved its headquarters out of North Adams. By 1992, the second owner had sold off all its divisions, and Sprague Electric had disappeared as an entity. Japanese corporations and their subsidiaries came to dominate the $20 billion global capacitor market, with China's industry share growing.

Completing the post-industrial transition, the current tenant of Sprague Electric's former 87 Marshall Street complex is the Massachusetts Museum of Contemporary Art, one of the world's largest contemporary art museums. Each day, visitors wander leisurely through galleries and studios enjoying the visual and performing arts on display in the building where thousands of workers once operated complex machinery for printing on cloth, and— later— sophisticated electronic component assembly equipment. Nonetheless, the transition in North Adams from an industrial to a post-industrial economy is proving to be a particularly difficult challenge.

ACKNOWLEDGEMENTS

John Storey, founder of Storey Publishing, read an early draft and provided useful and critical feedback on what is needed to create a publishable manuscript. So did William D. Middleton, III, coauthor with his father of the 2009 Indiana Press biography of my grandfather, Frank Julian Sprague.

Then, mired down and needing a fresh input, in the spring of 2012, I contracted with my son John to be my editor. An Amherst College English major and with an MA in Comparative Religion from Vermont College, John immediately began to attack discrepancies in structure, grammar, and format. Even more important, he challenged the objectivity of some of my views on labor relations and North Adams job losses and the resulting debate was educational and important for both of us. His input has helped create a far superior manuscript and in the process made us closer as father and son.

Throughout the research stage, my assistant Jean Lee, offered invaluable support in researching articles related to the history of Sprague Electric in local newspapers such as the *North Adams Transcript* and *Berkshire Eagle*. Similarly, Fred Windover, chief legal officer in the latter days of Sprague Electric, provided crucial input on copyright issues, and has been a key interface in helping obtain needed information from Penn Central.

Williams College has been an important resource throughout, including encouragement and support by Williams College President emeritus and good friend Francis Oakley, while current President Adam Falk has offered his views on the role of the college in local economic development. Geology Professor Reinhard ("Bud") Wobus provided much of the geological history, while History Professor Robert Dalzell referred me to Williams College Senior Honors Theses by Carin Cole, and Anthony Parise, which provided little-known and important historical background. Robert Volz and Wayne Hammond of the Williams College Chapin Library, along with archivist Sylvia K. Brown, not only made these theses available for study, but also recommended other sources of information. Finally, a recent 2012 Williams College Honors Thesis about Sprague Electric, *The Mark of Reliability* by Alison Pincus, made me realize how much local bitterness and misinformation related to Sprague Electric's mid-1980s departure from North Adams still remains.

Like Williams College, Massachusetts College of Liberal Arts has been another important historical resource. Linda Kaufman and Susan Denault at the MCLA Freel Library, identified and made available such sources as W. F. Spear's 1885 *History of North Adams* and Timothy Coogan's excellent NYU Ph.D. thesis. While retired Professor Maynard Seider and I have often clashed philosophically on labor relations issues, he provided me valuable information about Sprague Electric that he had been able to obtain from the Penn Central archives, as well as an incomplete manuscript he is currently writing. We remain friends.

Another important resource has been the North Adams Historical Society. Here Lorraine Maloney, some years ago, and now Charles Cahoon and Gene Carlson have been invaluable resources. The Museum is an important resource, and needs far more support than it is getting locally.

I also want to thank Denis Zogbi and Paumanok Publications, Inc. for allowing me to use selective data from *Passive Component Market Outlook, 2008 – 2013*. Since the Electronics Industry Association ceased operations in 2011, Paumanok is the only reliable source of such information.

For me the most interesting sources of information have been personal interviews, either by telephone or wherever possible in person. Especially in the case of former Sprague employees and associates, it has also been a way of renewing many old friendships, if only fleetingly. From Sprague Electric Don McGuiness, Bill O'Connor, Pete Loconto, Dick Morrison, Bob Milewski, and especially Dennis Fitzgerald (President of Allegro Microsystems) corrected my memory and filled in many missing gaps as we reminisced about the "good old days". I talked on the phone with former Sprague Technologies CEO, Ed Kosnik, and lunched with "Pug" Winokur, my last boss at Penn Central. "Pug" provided a view of Sprague Electric from the Penn Central side, which was both informative and sobering.

Local movers-and-shakers, including former Mayor John Barrett, current Mayor Richard Alcombright, John DeRosa and MCLA President Mary Grant have provided their views on a revitalized North Adams with an economy based primarily on the arts, tourism, and education, while local business heads and entrepreneurs Pam Art, Bo Peabody, Matt Harris, Bob McGill, Osmin Alvarez, Malcolm Smith, and Patrick Brennan provided direct experience and ideas on different business models to complement the arts and tourism. From Williams College and C^{3}D, Stephen Sheppard provided detailed analysis on what the Massachusetts Museum of Contemporary Art has contributed to the North Adams economy, and what still is needed to be done. Mark Maloy of the Berkshire Regional Planning Commission provided an even more sobering view of the economic future of the Northern Berkshires if current trends continue.

Massachusetts Museum of Contemporary Art Director, Joe Thompson, provided important editorial input. Along with MCLA, the continuing success of the museum under his leadership is by far the most important bright spot in the otherwise current cloudy economic environment of North Adams and, excepting Williamstown with its Williams College and Clark Museum, the rest of Berkshire County. Joe and I have also been in a several years discussion on an expanded companion publication to this book, titled *87 Marshall Street.*

Local historians Joe Manning and Paul Marino have offered their own thoughts on the past and the future of the Northern Berkshires.

Finally Robert Colburn of the IEEE History Center provided invaluable input to help make a somewhat rambling manuscript much more concise.

If I have overlooked anyone, please forgive what can only be called a "senior moment".

John L. Sprague
Williamstown, MA

I: SPRAGUE SPECIALTIES COMPANY

*The Sprague house in Quincy, Massachusetts.
(Author's collection)*

The Sprague Electric Company began as the Sprague Specialties Company in 1926 in Quincy, Massachusetts where— working part time in their kitchen—a young couple began building by hand the first product of a business that grew into one of the world's leading electronic component manufacturers. Sprague Specialties soon outgrew its Quincy location, and the company moved to North Adams in 1929. During World War II, it occupied the massive former textile complex of the Arnold Print Works at 87 Marshall Street and, like Arnold Print before it, eventually dominated the local economy. Sprague's world superiority in passive electronic components began to fade in the 1970s, and by 1990 Sprague Electric had been acquired twice and disappeared as an entity.

The first company by the name of the Sprague Electric Company was founded by electrical inventor Frank J. Sprague, who perhaps by chance lived his boyhood in North Adams and was undoubtedly the source of the skills of his three sons, Desmond, Robert, and Julian. He had considerable influence on all three, but often in more subtle than obvious ways.[1]

Robert Chapman ("R. C.") Sprague

Robert C. Sprague was born in New York City on 3 August 1900. He grew up to be of medium height, sturdy and well built (he was a good athlete throughout his life and still skiing in his '80s), with a square face dominated by a slightly curved nose. Behind his eyes lurked a certain aura of authority, and no matter what Robert Sprague became involved in, he inevitably ended up in charge. There is little information on his early life except from an extraordinary interview of Robert Sprague by his grandson, David, when David was a student at Williams College.[2] ("David Report" 1989 John L. Sprague personal papers)

Robert Sprague begins: "Originally I went to the Collegiate School in New York which was a very fine private school close to where we lived on 71st Street. Then my family sent me away for a year to the Curtiss School. It wasn't a good experience because of their demerit system which involved crushing rock. When my father came down to bring me home at the end of the school year, he had to stay an extra day for me to work off my demerits and my hands were pretty well bloodied from broken blisters. He didn't think it was an appropriate school from which to get an education. Neither did I.

"I went to Hotchkiss, Class of 1918, as a sophomore (I don't think they were called sophomores) [Editor's note: Hotchkiss uses the terms "Preps," "Lower Mids," "Upper Mids," and "Seniors."] and at that time if you were a Hotchkiss boy you were supposed to go to Yale. But the world situation was fairly ominous and my father talked a little about the possibility of my going to the Naval Academy instead. He was able to get me an appointment provided I could pass an exam, and I did. So I left Hotchkiss during my junior year and went to the Naval Academy when I was only sixteen. I entered as a member of the Class of 1921A, but because of the war in Europe we were accelerated to graduate in 1920. Then because of my high grades I was accelerated another year and actually graduated in 1919 at the age of nineteen." (No mean feat, no matter how brilliant the student)

Mishipman Robert C. Sprague. (Author's collection)

When asked how he was influenced by his father (Frank Sprague), Robert Sprague replied, "When I was growing up, he had been a leader in the electrical industry for twenty-five years, and I wasn't very conscious of this. Certainly in my early days I wasn't impressed with the fact that he was considered worldwide as a genius in this area. He was just my father. But he did have an experimental machine shop in New York and I spent a great deal of time down there learning how to use some of the tools, like lathes."

Robert Sprague met his wife, Florence Antoinette van Zelm after he left Hotchkiss for the Naval Academy. Florence's parents had a summer home on Twin Lakes, Connecticut, some fifteen miles from the Frank Sprague home in Sharon, and Robert Sprague's Hotchkiss roommate, Bob Lesher, had dated her. She was described as a wonderful dancer and "just about the most attractive girl" at a Hotchkiss dance that Robert Sprague had missed. Sprague fretted about how he could meet her until his younger brother, Julian, advised "why don't you just call her?" So Robert did, using the ruse that they had met at the Hotchkiss dance, which of course they hadn't. Initially she responded coolly, but—encouraged by her mother, a wonderfully loving and engaging woman—Florence invited him to visit her the following day. Sunday afternoon, he arrived in full-dress Navy whites, complete with white gloves, driving a modified Pierce Arrow 66. Robert would brag later, "She never had a chance." But anyone knowing "the van Zelm girl" would believe just the opposite was true. Regardless, that afternoon the romance began, a romance that would last another sixty-seven years.

van Zelm family photographs show Florence as a lovely slim young woman with an oval face, short bobbed brown hair and blue-grey eyes that seem to look right into your soul. In every one of these photographs she has a serious expression, except for one in which she is flashing a blazing smile in Robert's direction. Throughout her life this seemingly reserved woman hid a fun-loving streak as well as an iron will, both of which she would need in the years ahead. She never went to college, but studied piano and voice at the New York Conservatory of Music and continued with the piano at the New England Conservatory of music. She was able to play complex classical piano for many years, until arthritic hands made that impossible.

Florence's mother, born Antoinette Gray Hyatt and known as "Nettie", was from a family who were among the early settlers of New England. She cared only about her daughter's happiness and encouraged the relationship. But Florence's father was another matter. Johannes Louis van Zelm had been born in Rotterdam and had moved to the United States when he was only four. When Robert and Florence met, van Zelm was a vice president of the New York Life Insurance and Trust Company (later the Bank of New York and Trust Company). He was also serving the Netherlands in different capacities (in 1938 he was awarded the Officers' Cross of the Order of the Orange-Nassau by Queen

Wilhelmina). Much colder and more austere in appearance than actuality, van Zelm adored his daughter and initially was opposed to Robert Sprague as a suitor. "Just another sailor with a girl in every port." He went so far as hiring a private detective to try to determine—apparently unsuccessfully—what an untrustworthy person Ensign Sprague was.

Shortly after Robert and Florence met, Robert left for his first cruise as a junior engineering officer on the battleship *USS Pennsylvania*, a cruise that took him to Havana and then through the Panama Canal to Lima, Peru. While he was gone, he and Florence wrote letters to each other almost daily, and it soon became apparent that they wanted to get married as soon as possible after the fleet returned. Because Florence's father was still opposed to having the young people see each other, Ensign Sprague took control of the situation. "I finally went down to see him at the bank and told him, as nicely as possible, that whether he liked it or not, we were going to get married and hoped that he and Mrs. van Zelm would come to the wedding," Robert Sprague recalled. "Faced with the inevitable, her father finally relented and since then my relations with him couldn't have been better."

Florence Antoinette van Zelm and Ensign Robert Chapman Sprague were married on 24 May 1921, following which he attended the U.S. Naval Post-Graduate School, receiving a BS degree in Engineering in 1922. Then having decided to switch from a career as a general sea duty line officer to Naval Architecture, Robert Sprague went to MIT where he earned an MS in 1924. Following this he joined the staff supervising the design and construction of the aircraft carrier *USS Lexington* in the Charlestown Naval Shipyard. The young couple settled in Quincy, Massachusetts where they would live for the next six years. There they would form the Sprague Specialties Co., Robert Sprague would resign from the Navy (in 1928), and Robert Chapman Sprague, Jr. and John Louis Sprague would be born.

Robert C. Sprague and Florence A. Sprague early 1920s. (Author's collection)

While living in Quincy in the 1920s, Florence was active in the theater, and in 1983 was cited by Governor Dukakis for her 'unflagging dedication to the cultural community of the Commonwealth"[3].

Robert Sprague called her "dolly" (for darling) and most people remember her addressing him as "dear", or occasionally more sternly "Robert!" if annoyed.

Sprague Specialties Company

Sprague Electric began in Quincy, Massachusetts in 1925 with Robert Sprague's invention of a tone control device. In the 1920s, radios were all the rage, but Ensign Sprague didn't like the sound of the one he owned, so he experimented with methods of improving it using a group of condensers (now called capacitors) of different values to vary the tone of the radio. The result was a tapped fixed-paper condenser, of a completely new design, and an accompanying switch that could be used to choose seven different capacity values between the output of the radio circuitry and the speaker. This condenser could therefore vary the tone of the radio. People who heard the result thought it greatly improved the performance of the radio, and suggested that a business could be built around it. In early 1925, a series of patents were filed on the device, as well as on the unique condenser that was its heart, and *Tone Control* was registered with the U.S. Patent Office.

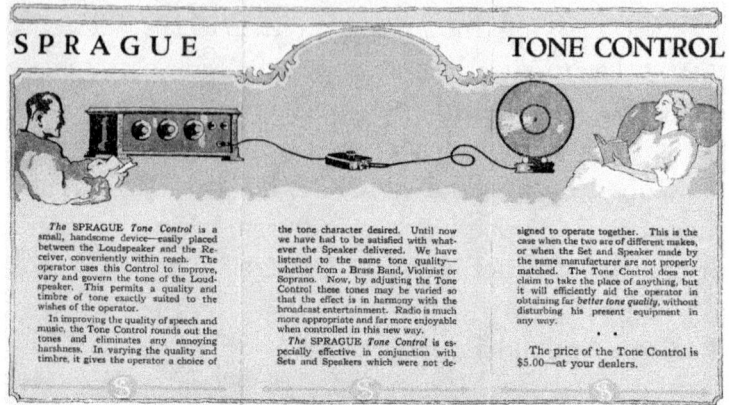

Tone Control flyer. (Author's collection)

A little over a year later, in June 1926, the Sprague Specialties Company was incorporated with an initial capitalization of $25,000, of which $3,200 came from the young couple's personal savings. Because Ensign Robert Sprague was still on active duty in the Charlestown Navy Yard, initially Sprague Specialties was strictly a moonlighting operation. Everything was done by Robert and Florence in their Quincy home, first in the kitchen and then in the basement. It is certain Florence Sprague hadn't bargained for simultaneously supporting a husband with two jobs, bringing up a young child (Robert C. Sprague, Jr. was born in 1923), doing most of the paperwork for a fledgling company and— in her spare time—helping manufacture a strange thing called a condenser, all in her mid twenties. She was relieved of the paperwork part of her job when the company's first paid employee, Miss Mary E. ("Molly") Avery joined them and assumed responsibility for correspondence, bookkeeping, and other clerical work. Molly also became personal secretary to Robert Sprague, a position she held for the next thirty-five years. Although the original plan had been to subcontract the manufacturing of the condensers and their assembly into the tone control, as well as the promotion and sales, "that didn't work out as planned and we had to take over both manufacturing and sales, without any experience in either." Even worse, no one seemed interested in purchasing the finished device, and by mid-November 1926 more than half of the seed capital was gone.

Robert Sprague's younger brother, Julian, had also joined the firm shortly after it was formed. His title was Production Manager, although like everyone else he initially was a jack of all trades. Even though he lacked his brother's formal engineering background and training, Julian was extremely bright and made a crucial early contribution when he concluded that the key asset owned by the company was the heart of the tone control, the small tapped paper condenser. He suggested making a single fixed condenser using the same basic paper/metal film construction as the multi-position unit. The resulting "Midget" was half the size and one eighth the weight of the standard mica condenser then offered by such established giants in the electronics field as Cornell-Dubilier. It had nearly equivalent electrical and life performance, and—because of the method of construction and materials—was much less expensive to manufacture. Following Julian's suggestion, the Tone Control was shelved, and using what was left of the initial capital, the company bet its future on the "Midget."

During its life, Sprague Electric manufactured capacitors using just about every different dielectric system available. Despite the success of the Midget®, even mica was no exception. While it has poor volumetric efficiency, it is extremely stable, chemically inert, has very low leakage, a high breakdown voltage, and can be used in high-temperature environments. A typical application today is for high power RF transmitters. For many years there were two specialty mica businesses in the North Adams Brown Street plant (and later Grafton, Wisconsin). One used pure slit natural mica sheet as the dielectric, while the second used Fabmica®, a material made of mica pieces in a synthetic carrier. Volumes were limited, but profits reasonable. For a number of years the resident genius was Harold Brafman, who joined Sprague in 1941 and by 1958 had more than thirty years' experience in the Mica field.

Conceptually, a condenser or capacitor (the more modern term) is an extremely simple device in which two— generally metal—conductors are separated by an insulator (dielectric). The dielectric can be air, paper, a synthetic film such as mylar, mica, metal oxides of aluminum or tantalum, or ceramic materials. Capacitors are used to store energy, and they block the flow of direct current while passing alternating current. The first was a "Leyden Jar" developed independently in 1745 in Germany and Holland, and a year later in England by John Bevis. Leyden Jars were glass jars surrounded by two pressed metal conductors, and their ability to store electric charge or capacitance was extremely low.

The electric energy storage capability ("capacitance") of a parallel plate capacitor is:

$$C = K'KA/d$$

where K' is a constant, K is the dielectric constant of the insulator, A is the area of the plates, and d is the thickness of the dielectric. Therefore, for a capacitor to be useful, it needs to have a high dielectric constant, lots of area, and as thin a dielectric as possible, as long as it still has satisfactory breakdown strength (so that a short circuit won't occur in the dielectric under the voltage used in the application). Over the years some exotic materials and construction techniques have been developed to satisfy these requirements, many of them by Sprague Electric. Commercial use of capacitors began in the late 1800s, first in telegraphy, then telephony, and finally radio. Today they are used everywhere. Use of stacked cleaved mica as a dielectric was first demonstrated in 1845 while early work on aluminum electrolytic capacitors began nine years later.[4] The superiority of the "Midget" came from much less expensive materials and construction (rolled metal film and paper versus stacked metal film and cleaved mica) and a considerably thinner dielectric.

Sprague Electric sent samples and promotional material to radio manufacturers, and the response was immediate and positive. Despite some concern about the long-term viability of a new and untested company, orders began to pour in, and the tiny firm had to scramble for additional manufacturing space, new machinery, and production workers.

"Sales in 1927 were $54,000 as against almost nothing the year before, and increased to $234,000 in 1928 with a net profit of $35,000. By the time it neared its third birthday in 1929, Sprague's employment had risen from the original two persons to a peak of 550 and its sales were running in excess of half a million dollars." (from a 6 November 1958 presentation that Robert Sprague made to the Newcomen Society).

Although Robert Sprague was both President and Treasurer, he was no longer working alone. In March 1927, Sprague Specialties was located at 1511 Hancock Street in Quincy. William J. Nolan was Clerk and, until his death in 1974, worked as the company's Chief Counsel, member of the board, and close confidant to Robert Sprague. Harry Kalker was also one of the earliest employees, joining the Company in 1926, and forming Sprague Products in 1933 to handle "consumer sales" to—for example—radio parts jobbers who used Sprague capacitors in repair of radio sets. Such retail sales grew to account for nearly one third of all of Sprague's electronic component sales.

Although he probably could ill afford it, another early stockholder was Frank J. Sprague. Although it appears that he exerted no direct influence on the running of the Company, and attended meetings only by proxy, his presence must certainly have been felt. By 1933, all three of Frank Sprague's sons and his son-in-law were employed by Sprague Specialties. Although Desmond was Frank Sprague's eldest son, after the failure of Frank Sprague's last two companies, Robert Sprague assumed responsibility for trying to help his siblings in any way he could. When Desmond joined in 1929 as "Plant Engineer," he was without a job. Bud Tucker, who was married to Althea, (Robert and Julian's sister), and who joined the sales force in 1933 at the height of the Depression, was probably likewise without employment. Robert Sprague later vowed that he would never allow his own family to face the same type of financial difficulties he had after the death of his father.

Sprague family circa 1932. Standing from left to right: Julian's wife Helene, Julian, Robert, Desmond, Desmond's wife Ruth, Althea's husband Bud Tucker. Seated left to right: Harriet, Frank, Althea. Missing: Robert's wife Florence. (Author's collection)

With the Sprague Specialties Company's retail sales growing and the company expanding, Bill Nolan, William R. Hurley, H. Fred Lalley, and Julian K. Sprague were elected early Directors by the Board, in addition to Robert Sprague. Over the next several years, Hurley and Julian became Vice Presidents. In April 1928, capitalization was increased, and regularly thereafter. In March 1929, the President's salary was set as $7,500 annually. Just three weeks before the New York Stock Exchange's collapse on "Black Tuesday" (29 October 1929), Sprague Specialties purchased the Beaver Mills from the North Adams Industrial Company for $42,000, with the funds derived from the sale of blocks of preferred and common stock to North Adams individuals. Extensive credit to finance the move and provide additional capital was obtained from Boston and North Adams Banks. In addition to the new plant, at the Annual Meeting (held March 3, 1930) there seemed only good news to report: earnings of $73,899 (before taxes, depreciation, and reserves) on sales of $576,534; net worth of $492,282; and a brand new aluminum electrolytic product line[5]. Revenues for 1930 were forecasted to exceed $1,000,000, and there were important new management additions in Sales, Engineering, Operations, and Research. In light of what was happening in the world economy, it seems impossible to understand such a rosy outlook, but as Robert Sprague continued, "We were young, optimistic, and certain that we had what it took to run a business."

In 1929, two management additions helped enhance the future of the fledgling company. After four years in ship construction at the Portsmouth Navy Yard, Carleton Shugg, a brilliant former naval officer and classmate of Robert Sprague's at both Annapolis and MIT, was hired as production manager and given complete responsibility for bringing the Beaver Mills into production. Shugg served in that capacity until late 1940 when—feeling he could best serve his country by returning to shipbuilding—he resigned to join the Cramp Shipbuilding Company in Philadelphia. Shugg had a remarkable career. After several more shipbuilding jobs he moved on to the Atomic Energy Commission[6] eventually becoming its General Manager. There he hired Hyman Rickover to lead the nuclear propulsion program. In 1951, Shugg joined the Electric Boat Division of General Dynamics and soon became Division President. While there he and Rickover began to build the United States' fleet of nuclear submarines, and, in the late 1950s with Admiral William Raborn, the first missile-equipped submarines. He retired in 1965 and lived until 1992.

Although Carl Shugg was still a young man when he was at Sprague Specialties, he was Robert Sprague's closest confidant. Shugg was a short, powerfully built, unsmiling man with a brilliant mind, a seemingly encyclopedic understanding of how to manufacture almost anything, and a person who always said exactly what he believed regardless of the consequences.

Throughout his business career, Robert Sprague always emphasized research and development as the most important way to feed the company with a continuous portfolio of new products. Several new developments, (the most important of which was the Aluminum Electrolytic Capacitor) that started in 1929, helped save the Company from almost certain bankruptcy. In 1929, Dr. Preston Robinson was hired to formalize and staff Sprague Specialties' first Research and Development Department. Dr. Robinson grew up in Brookline, Massachusetts, received his BS in 1922 and MS in 1923, both from MIT, and then his PhD from the University of California in 1925. He returned east to work in the Guggenheim Research Laboratories where he met Frank J. Sprague and soon afterward joined Sprague Specialties in Quincy as Chief Technical Officer. He was then elected VP Research and Engineering, and in 1932 he joined the BOD, a position he maintained until 1967. In 1953 his administrative responsibilities were assumed by Dr. Wilbur Lazier and Robinson became a senior technical consultant, a position he maintained until his death in 1973. Measured by number of patents, he was Sprague's most prolific scientist with one hundred and fourteen U.S. patents, the most important of which was the basic patent for the solid-electrolyte capacitor. On 25 May 1965, he received special recognition for receipt of Sprague Electric's 500th patent. Still filing patents in his late sixties, he died in 1973.

In the town of Quincy, there wasn't enough room for the forecasted level of employment. The existing space was in violation of the local sanitation laws, and the addition of the required extra toilets would have required expensive renovation.

"After considering locating in some of the larger metropolitan areas around Boston such as Lowell, and further west, Holyoke, we decided we should also look at the far western part of the state where we had heard there was plenty of available factory space and labor in Adams. So two of us took Route 2 over the Mohawk Trail and arrived in North Adams in the early afternoon. But unable to find any road signs to Adams, we stopped to ask directions from a man standing on one of the corners. Looking us over carefully, he asked why we wanted to go there. On learning the reason for the trip, the man (believed to be either Frank Bond or George B. Flood) declared, 'hey, you don't need to go to Adams, we have everything you need right here in North Adams."

The 3 October 1929 *North Adams Transcript* trumpeted, "Electrical Industry Employing 1,000 to Locate Here; Sprague Specialties Co. of Quincy, Massachusetts Selects North Adams in Which to Expand; Beaver Mills Bought; $100,000 Is Raised by Business Men of North Adams." The article continued with a brief history of the Company, a description of the competitive battle with other Massachusetts cities such as Lowell and Holyoke, and there was a long accompanying article about Frank J. Sprague.

"The North Adams renovated plant began operations on 1 April 1930 (with twenty-five young women who had been through a two week training program), and by mid-year employment reached two hundred, and was forecasted to reach six hundred to seven hundred by year-end 1930. In June 1931 the Sales Office and 'Experimental Laboratory' moved from Quincy to Beaver Street and the Sprague Company had become a one hundred percent local North Adams industry. At the end of 1930 our optimism seemed justified. Sales had increased almost fifty percent over 1929, although we were beginning to see a disturbing downward trend in the ratio of net profit over sales. Next year, we told ourselves, we will tighten up."

The Beaver Street plant contained a huge manufacturing floor with rows of benches where women workers sat side by side operating sophisticated equipment, some rolling capacitor sections while others assembled the sections into round aluminum cans. The area was bright, cluttered but clean, and there were three distinct sound levels: from fans attempting to cool the summer heat, from the clattering of the equipment, and from conversation between the operators. Somehow they were able to manufacture parts at high speed while at the same time they talked about their homes and daily life. Because the factory was in the Beaver section of North Adams, which had originally been settled by French Canadians, at times the conversation was French. They were hired and trained locally and most came from families who had been part of Arnold Print.

The "etch house" was another story. It was a dark, smelly place filled with long machines in which aluminum foil was continuously dipped into cauldrons of bubbling acid. In some of the machines, the surface area of the foil was pitted to increase its surface area by applying a positive direct current voltage to the roll of foil as it passed through a bath of—typically—hydrochloric acid. In other machines, an insulating aluminum oxide film (the dielectric of the finished capacitor) was grown by making the now pitted aluminum film the anode in an electrolyte, such as aqueous boric acid and sodium borate. The equipment in the etch house was operated by men. In general, at Sprague the assemblers were women, while management, engineers, technicians, and those operating complex equipment, such as for etching and formation, were men.

Especially in capacitors, almost all of Sprague's manufacturing machinery was made in-house starting with an engineering design. In the February 1939, *Log* (a newsy internal Company periodical), there is a wonderful picture of a Sprague engineer leaning over part of an electrolytic formation machine. The caption states that "This is the fifth machine developed for this operation since it started eight years ago, the first being a standard clothes wringer perched on the edge of a wooden tank!"

An impressive 1929 list of customers included Burroughs, Columbia Radio Corp., GE, Magnavox, Raytheon, and Westinghouse. Soon Philco would be added to the list and would become Sprague's largest customer, holding that distinction until 1950s when IBM took the top customer slot.

July 1931 saw the introduction of an unusual invention by Robert and his brother, Julian. Called the Visivox, it was a complete home entertainment system that included a 16-mm movie projector, a synchronized record player, an amplifier/radio, and a built-in speaker. There were even plans for a circulating film library. However, the project was quickly dropped because of the highly negative reaction of some of Sprague's customers, especially Philco, who objected to having one of its component suppliers also a competitor. One of the few surviving Visivox models is often on display in the lobby of the Massachusetts Museum of Contemporary Art at 87 Marshall Street.

In 1931 sales exceeded $1 million for the first time, but soon the roof fell in. Despite record sales, expenses were now out of control as "our informal management systems had broken down and our expenditures were significantly above our receipts, (causing) a net loss of truly staggering proportions ($295,000 on sales of $1,175,000)".

"Most of the new research-based enterprises in New England, such as Sprague Specialties, were created and operated by people with scientific and technical rather than general business backgrounds," Robert Sprague recalled. "In 1930 most of our business experience was limited to supplying a product greatly in demand to a market which seemed to be growing by geometric progression. So the New York consulting firm of Stevenson, Jordan and Harrison was brought in to not only modernize the Company's accounting and operational controls, but also study plant layout and compensation systems. Inventories were slashed and, among other draconian measures, all product and process research was temporarily eliminated as well as any new equipment and facilities expenditures."

The Sprague Electric Company also needed to obtain extensions on its bank loans and from other major creditors of at least two years. Important local help came to the rescue. At the request of the bank creditors, George B. Flood joined the company to look at its finances, first without pay as an informal observer and then as full-time Treasurer, replacing Robert Sprague in this capacity. With extensive prior business experience as treasurer of Arnold Print Works, he became a key member of the management team and would remain in that same capacity at Sprague until 1953 when Robert Sprague again assumed the treasury functions. Flood's son-in-law, Neal Welch, joined the sales organization in 1932 where he first worked in inside sales. Rising through the organization, he succeeded Julian Sprague as vice president of sales after Julian's untimely death in 1960, and Robert Sprague as CEO in 1971.

The rigid cost controls worked, at least as far as short-term financial stability was concerned, although long-term viability was severely threatened in the process. Although 1932 sales were only $671,000 (fifty-seven percent of the prior year) the loss dropped to $32,800, and from then on—despite a number of economic cycles—the Company had an enviable record of continuous profitability until 1968. By 1936, not only had Sprague Specialties survived, it had paid off nearly all of its $800,000 of deferred liabilities, had nearly $250,000 cash in the bank, and it reported record sales of $2,902,000 and profits of $284,000. In early February 1937 the Company purchased the former Hoosac Worsted Mill plant on Brown Street for further expansion and to protect the value of the headquarters location and the reliability of the products manufactured there. For example if a cement company had bought the mill, cement dust in the air would have been a major problem affecting the reliability of electrolytic capacitors manufactured in 87 Marshall Street.

Although between 1933 and 1936 sales had nearly tripled, this was mostly because of new products that had been developed prior to 1932. Further new product development had stopped in 1932, and the results of the stoppage were about to be felt. Product life cycles vary widely in different electronic component families, but in capacitors the gestation period for a truly new product, from research to billable sales, can easily exceed three years. With a greatly improved financial base, in 1936 Sprague reactivated its dormant research and development activities and began to hire an extremely competent group of scientists and engineers to rebuild its laboratories. The results of this effort would not begin to appear until the end of the decade, luckily just in time to respond to Pearl Harbor.

During the Great Depression, all of Sprague's competitors, such as Cornell-Dubilier and P. R. Mallory Co., had also been forced to curtail expenses, but their research and development had not been reduced to nearly the same degree as Sprague Specialties. Robert Sprague recalled that, "by 1937, competition was finding ways of equaling or improving much of what we had to sell. Whipsawed by competition on old products, and stymied by the laboratory-to-production time lag on new ones, (despite an impressive customer base that included Philco, Delco Radio, Emerson Electric, Ford Motor Company, and Westinghouse) our 1938 sales were only two-thirds of those for 1936 and profits had slipped from eleven percent to under three percent."

Since its formation, Sprague Electric had developed a well-earned reputation for superior quality and reliability. As one example, on a mid-1930s sales call to Philco, Robert Sprague sat quietly in front of the purchasing manager as Sprague's competitors gave their pitches with their units lined up on the desktop blotter. The capacitors in question were aluminum electrolytics used in radio power supplies, and corrosive electrolyte leakage was a major industry problem. Finally the buyer turned and said, "Well, Mr. Sprague, you've been pretty quiet. What do you have to say?"

Robert Sprague leaned forward and picked up the Sprague unit, noting, "Ours don't leak."

The blotter was dry only where the Sprague unit had been sitting. Around the location of every other competitive unit there was a gradually spreading ring of electrolyte.

Still, as the European storm clouds thickened, Sprague Specialties continued to struggle. Also for the first time, relations between labor and management became contentious, and there was a one-day strike in 1936. In the mid-1930s, the United Electrical, Radio, and Machine Workers of America (or UE), an affiliate of the CIO, began to organize production workers and machinists at equipment companies such as GE, Westinghouse, RCA, Philco, and others until by 1943 their contracts covered nearly 600,000 workers in the electrical and radio industries.[7]

On the other hand, during most of the 1930s labor strife at Sprague Specialties had been relatively quiet. In no small part this was because the production workers were represented by an independent union, first the Independent Condenser Workers #1 (ICW # 1) and then ICW #2. The power of the ICW lay in its six-man Grievance Board which usually met monthly informally with management to discuss and resolve grievances. Labor contracts were negotiated on an annual basis, allowing for incremental, rather than major, changes in wages and benefits. Sprague Electric also worked hard to keep a family-like "we are all in this together" relationship with its work force. Working conditions were good, management had an open door policy, and during the lean times when there were difficulties meeting payroll, the company offered inexpensive $0.50 stock to its employees. Although the shares held no real value at the time, for those employees who kept the stock, there was a very hefty return years later, especially in the early to mid-1960s. Following a mid-January 1938 talk by Carl Shugg to the ICW #1 on the seriousness of the Company's financial situation, the union agreed to a temporary ten percent wage reduction which was later rescinded as business improved.

In September 1941 there was a more serious nine-day wildcat strike with workers demanding a fifteen percent wage increase, although ICW #2 had already negotiated a five percent increase to run until March 1942. Behind these demands were efforts by the UE, which represented General Electric's large manufacturing complex in nearby Pittsfield, to gain representation rights for the Sprague workers. These efforts failed, and although the company did make concessions, the 24 September 1941 settlement largely favored management. Labor unrest did not cease, and the UE kept up its efforts as company financial performance continued to improve and employment approached 3000, making Sprague Specialties an increasingly attractive plum to pick. The next serious challenge came right after VJ-day and led to an extremely contentious six-week strike.

Reviewing a dismal 1938 in the January 1939 issue of the *Log,* Robert Sprague reported the introduction of two important new product families: a line of "Atom" small etched dry electrolytic capacitors and—for the first time— a fixed resistor product line. This was the beginning of a corporate strategy, which World War II would greatly accelerate, to become a full-service supplier of many different types of electronic components beyond capacitors. A year later, Robert Sprague reported a twenty percent sales increase in 1939 over 1938, production of a record fifty million units, employment at a record 1300, and potential new markets in television and FM radio. The increasing importance of Sprague Specialties to the local North Adams community was also seen when in 1939 there were more than eighty different North Adams-based suppliers of goods and services to the Company.

Although sales and profits dipped slightly in 1940, it became evident that the United States would eventually be drawn into what at that time was still a European war, and the switch to a wartime economy began. In July 1940, Sprague Specialties and the Wall-Streeter Shoe Company jointly bid successfully against fifty-six other firms on a $172,000 trial government contract to manufacture gas masks for possible civilian deployment. A shoe company might seem like a strange partner, but when it was learned that stitching was an important part of the manufacturing process, Robert Sprague turned to Wall-Streeter President and friend James E. Wall for help. The plant at Brown Street was renovated and eventually set up as a special facility to manufacture military products, including a test laboratory where products were stress tortured to determine life expectancy. Gas mask production began in Brown Street in January, 1941 by the Wall-Streeter "Fitting Division" under the supervision of Mitchell F. Nejame. The masks themselves were claustrophobic, had an awful smell, and fortunately were never needed.

I: SPRAGUE SPECIALTIES COMPANY

Toward the end of 1940, Carleton Shugg resigned as production manager to return to shipbuilding. Although he would be missed, his excellent replacement was Amos Carey, who came from RCA with extensive experience in the manufacture of U.S. Government and military equipment. Carey's responsibilities quickly expanded as first Brown Street, and then Marshall Street, cranked up to manufacture increasingly complex electronic components and materials for wartime applications. It was a tumultuous period of extraordinary growth followed by temporary decline from which Sprague Electric emerged as a much stronger company well on its way to becoming a worldwide powerhouse in the electronic component industry.

Because there were no Annual Reports until 1945, and because much of the work was classified, piecing together the war years at Sprague Electric requires a variety of sources, including the 1958 Newcomen Society report, internal company documents, a special 7 December 1960 issue of the *Transcript*, other *Transcript* articles published during World War II, a Williams College thesis by Raymond Bliss, and issues of the *Log* during that same period.

The contributions that Sprague Specialties/Sprague Electric made to the war effort and what it accomplished between 1940 and 1945 were impressive. In 1940, sales were slightly more than $2 million, there were only some 1300 employees in one plant (Beaver Street), and the primary product line was aluminum capacitors for commercial radio power supplies. When the war ended in 1945, sales had grown eight-fold, employment was now spread across three plants, and had peaked at more than 3000, and the Company had made crucial contributions to the war effort, including components used in microwave airborne search radar, the VT proximity fuze, and in atomic weaponry. It had also manufactured gas masks and incendiary bombs and— in the process—been awarded five Army-Navy "E" Awards and one Navy Bureau of Ordinance "E" Award.

The July 1940 gas mask trial order was awarded to Sprague Specialties/Wall Streeter as the low bidder and production began initially in Brown Street and then in mid-1943 it was moved to leased space in the Arnold Print Works complex on Marshall Street. Multiple add-on orders continued until production was terminated in April 1944.

A 27 October 1940 issue of the *North Adams Transcript* reported that Sprague Specialties had submitted bids on components for a new type of incendiary bomb. There was little other information because Army Regulations forbade comment on the work. Less than a month later, the *Transcript* carried another article saying it was rumored that the company had received an $800,000 order to make bomb parts. The company again refused comment, and after that there was mostly silence except for an article on defense programs in the May 1941 issue of *The Log*. It reported production of various types of capacitors for army and navy communications gear, and stated that Brown Street was being specifically tooled as a "Defense Plant."

Although gas mask and incendiary bomb orders in 1940 and 1941 anticipated involvement in what was still primarily a European conflict, the United States was not yet completely committed, and pockets of isolationism existed across the country. In Robert Sprague's Christmas message in the December 1941 *Log* (obviously written before 7 December), he gave a mixed message: "To preserve our country and to put down surely those who would replace our Bible with 'Mein Kampf,' we must hold fast to our belief in the need for good will among men. Let us, therefore, celebrate this holiday season with good cheer and keep the hearts of our children bright with the spirit of Christmas. And now looking forward to 1942, it is not going to be easy. It will be hard. So let us all together put our shoulders to the wheel in this common cause and PUSH – ON – TO – VICTORY in 1942, 1943 or however long it takes."

World War II

At 7:55 AM on Sunday, 7 December 1941 nearly two hundred Japanese torpedo bombers, dive bombers and fighters, which had been launched undetected from the Japanese fleet some two hundred and seventy-five miles north of Hawaii, swept over Pearl Harbor on the Hawaiian Island of Oahu where the bulk of the United States' Pacific Fleet lay.

Even as the United States was transformed into a full wartime economy, the Sprague *Log* continued its almost sunny outlook on the progress toward victory and peace. It featured less news on sports and social events, less about what the company was actually doing to support the war effort, more letters of congratulations from customers on delivery of war-related components, an increased emphasis on how to "do better" in wartime, and continual urging in issue after issue to purchase war bonds. The *Log* had always been about people and marriages and births, and here the tone changed with time. From the beginning women had made up the bulk of the manufacturing employees at Sprague Specialties and similar companies. This became even more true as husbands, sons, brothers, and lovers went off to the Pacific, Atlantic, and Europe, some to return cruelly wounded, and some never to return at all. There was more and more news from the front, more censored letters about how the writer's own personal war was going, and increasing reports of fateful "gold star" telegrams, letters, or personal visits.

A 1 June 1942 telegram from Major General Porter, Chief of the Chemical Warfare service, reported that incendiary bombs (actually Sprague made the bomb casings which were filled with the explosive elsewhere) manufactured in the Brown Street plant had been dropped over Tokyo and several other Japanese cities during Colonel James Doolittle's 18 April 1942 carrier raid from *USS Hornet*.

Employment continued to grow as electronic components and other war-related products poured out of Sprague Specialties and from its six subcontractors. These included The Rock of Ages Capacitor Corporation in Barre, Vermont, which specialized in manufacturing film capacitors and which was eventually purchased by Sprague Electric.

As reported in a "Sprague Day" special 2 April 1943 edition of the *Transcript*, Brigadier General A. A. Farmer and Captain J. S. Evans of the War and Navy Departments came to North Adams to present Sprague Specialties with the coveted Army-Navy "E" Award "for high achievement in war production." The ceremony was held at Brown Street and, in addition to a host of dignitaries including Massachusetts Governor Leverett Saltonstall, all Sprague employees (who now numbered more than 3000 out of a North Adams population of roughly 22,000) were in attendance. Following the ceremony, each was presented with an "E" pin and certificate. In an emotional acceptance speech, Robert Sprague described some of the award's background:

"The idea of an 'E' Award was originally conceived by the U. S. Navy in 1906, and first awarded for excellence in gunnery, and later for excellence in engineering and communications. From that date any ship with a white 'E' painted on a turret or stack has been considered outstanding in her class. Since the start of the War, the significance of the 'E' has been broadened to include recognition for outstanding performance on the production front. But we must redouble our efforts and at the end of six months if we again make the grade we can add a star to each of these flags (that now fly over the Beaver and Brown Street plants)." Sprague plants would receive four more stars. During World War II fewer than five percent of all war plants received an "E" award; such honors were important, and they served the reputation and subsequent growth of the Sprague Company.

1943 Army-Navy E Award (North Adams Historical Society)

Gas Masks and incendiary bombs were obviously important war items, but *Transcript* articles hinted at even more significant contributions.[8] One of the most important was design and production of special capacitors for the U. S. Navy's VT proximity fuze. Located in the nose of a projectile, the fuze was a rugged and complex miniature five-vacuum-tube radio and receiving device which caused the shell to explode if it passed within a predetermined distance—say seventy feet—of the target, rather than on contact or by a preset timing device. Research Director Dr. Preston Robinson headed the design team; production began in September 1942, and the first successful hit was made on 5 January 1943 when the cruiser *USS Helena* shot down a Japanese plane using a VT proximity fuzed shell. The device played a dominant role in combating Japanese kamikaze attacks late in the war, while the Army version, which had been held in reserve until the crucial Battle of the Bulge, was credited in helping blunt the German offensive and saving thousands of American lives. At its peak, the Sprague Company employed 2,400 on the project.

On 1 October 1945 Robert Sprague announced that, in addition to the 5 Army-Navy "E" Awards already received, the company had been awarded the coveted Navy Bureau of Ordinance "E" Award for its role in development and manufacturing of the VT fuze, one of only thirty-six firms so honored out of the thousand that participated in the program.[9]

Finally, as part of the super-secret Manhattan Project (the code name at Sprague was "Manhattan Square"), a team—again led by Dr. Preston Robinson—designed the Sprague Company special energy storage capacitors that were used in the trigger mechanisms of the first atomic bomb exploded at Alamogordo, New Mexico, and in the nuclear weapons that were subsequently dropped over Hiroshima and Nagasaki, Japan, bringing World War II to an abrupt and horrifying end. Following the end of World War II, Sprague Electric continued to supply them for the United States' growing production of atomic and nuclear weapons.

On 24 April 1944 the company officially changed its name from the Sprague Specialties to the Sprague Electric Company, a name felt to be more representative of its businesses. Sprague Electric Company had also been the name of Frank. J. Sprague's most important company, the one that was sold to GE in 1902. To allow room for further expansion, in August 1944 Sprague Electric Company purchased the former Arnold Print Works complex on Marshall Street in North Adams from the Beacon Realty and Trading Company, which had purchased the property in 1942 when the Works went bankrupt and ceased operations.[10]

As the wars in Europe and the Pacific moved toward bloody conclusions, detailed planning for the transition, or reconversion as it was called, began at Sprague Electric and throughout the country. Prices had been fixed by law since 1941; increases in these after the war would have to be negotiated. Almost everything Sprague made was related to the war effort, therefore major cancellations were expected. Friction with labor began to intensify. As negotiations for the fall 1944 labor contracts started, Local

#249 of the Congress of Industrial Organizations' United Electrical Radio and Machine Workers of America (UE), led by Gerry Steinberg, announced it would once more seek representation of the Sprague workforce. A new and powerful force had joined Sprague's ICW#2. William Stackpole, who came from the General Electric plant in Pittsfield and was violently opposed to the UE, moved to Sprague Electric in the middle of the fray (two months later he became Chairman of the powerful ICW#2 Grievance Committee, a position he would hold for more than twenty years). He attacked the UE as being both useless and communist-leaning (a common tactic nationwide at the time) and, when the election was held the end of August, again the UE was defeated.[11] But labor problems were not yet over.

With World War II officially ended on VJ-Day, 22 August 1945, "reconversion" began at Sprague Electric and across the United States. Eighty percent, or some $8.5 million, of Sprague Electric's outstanding orders were cancelled. Sprague immediately terminated its six subcontractors and moved to lay off workers, institute a reduced work week, and temporarily close Marshall Street and concentrate operations in the Beaver and Brown Street plants. Prior to VJ-Day, total Sprague employment in North Adams was 2600, a number that briefly dropped by more than 1000 in less than a month. Employment recovered as new orders came in, but only after significant damage had already been done.

Just before the war ended, labor problems intensified. The 4 May 1945 issue of the *Transcript* reported that "1,523 at Sprague's In Favor of Strike" if a satisfactory new contract were not reached. In October there was a two-day walkout and an emboldened ICW #2, led by William Stackpole, demanded an across-the-board thirty percent (later reduced to twenty-two percent) wage increase to make up for lost income caused by the shortened work weeks. A bitter six-week strike followed, settled on 16 December with terms that clearly favored the company. Because the transition from military to civilian orders was in its early stages, although the workers lost income during the six weeks, there was little impact of lost production to customers. Robert Sprague summarized the result, "Like most employers in the country, (following the end of the War) we found ourselves faced with labor unrest. We were unable to come to an agreement on a wage increase and a strike was called on November 1. After prolonged negotiations an agreement was reached and the employees returned to work on December 17."

Although there were later periods of unrest—and although the UE, as well as several other national unions, continued their representation efforts—with the exception of a five-and-a-half month strike in the latter half of 1949 by some thirty machinists (who had affiliated with the International Association of Machinists in February 1949), there would not be another strike of production, office, and maintenance workers at Sprague until 1970. Several historians, including Bliss, have concluded that this was because of the weakness of both the UE and ICW #2.[12] There is, however, another obvious possible conclusion that, during the extended period of growth and prosperity at Sprague following the end of World War II, ICW #2 was able to represent the workers needs and demands fairly, and that Sprague Electric was a good place to work.

Table 1: Net Sales Billed and Profit After Tax, 1941 – 1946

Year	NSB	PAT
1941	$4,796 M	$209 K
1942	7,373 M	208 K
1943	14,469 M	548 K
1944	20,801 M	872 K
1945	16,724 M	654 K
1946	10,767 M	720 K

After being in business since 1926, Sprague Electric issued its first Annual Report to Stockholders for the Year Ended 31 December 1945. Despite reconversion difficulties, net sales billed totaled $16,724,298, profit after tax $653,913, and cash on hand $2,113,270, a very good start for the next growth period. The Annual Report also noted that five hundred and fifty-one Sprague employees served in the armed services, of which eighteen "made the supreme sacrifice." More than ninety percent of those so far discharged from the military had already returned to work at Sprague.

Pent-up demand for radios, household appliances, automobiles and other consumer products was expected to fuel the need for electronic components such as capacitors, and it was expected that equipment manufacturers would buy from long established suppliers (such as Sprague) who are known for high quality and on-time delivery. Not anticipated in this forecast were the renewed military requirements that soon resulted from the Korean War, the increasing tensions between the United States and the USSR as the Cold War heated up, and the introduction of television. Sprague Electric was poised to begin a long and productive growth period.

II: SPRAGUE ELECTRIC JOINS THE SOLID-STATE REVOLUTION

Sprague Electric's wartime sales peaked in 1944 at $20.8 million, then dropped steadily for the next three years as the United States converted from a wartime to peace-time economy. Revenues in 1947 reached a low point of $10.5 million, although the company was still able to eke out a small profit. Employment had also dropped to 2100. But the decline was short lived as demand for new consumer products such as television surged. When the North Koreans, backed by China and Russia, swarmed across the 38th Parallel on 25 June 1950, and the Cold War began in earnest, demand for electronic components increased even more. By 1953, revenues had grown to $46.8 million, profits to $2.89 million, and employment reached an all-time high of 5,500. (see Table 2 in Appendix 1)

In order to meet increasing demand in both the consumer and military markets, Sprague Electric continued to fill its three main North Adams plants on Marshall, Brown, and Beaver Street. Being old mill buildings, they offered large spaces that could be made attractive with careful layouts and artfully placed paint and color-coding of the walls and overhead pipes. They also offered lots of space within which to build rooms where special environmental and/or electrical control was necessary. On the other hand, they were very expensive to heat, or cool, and to maintain. The annual cost to maintain and repair Marshall Street's windows alone was close to $50,000. Although well-suited for engineering labs and office space, Marshall Street—with its twenty-six buildings interconnected by outside walkways, tunnels, and overhead bridges—was less adaptable for manufacturing. Added to the physical plant difficulties, Sprague Electric management was concerned about saturating the local labor market, and potential union problems. It launched a branch plant program in the neighboring states of Vermont (Barre and Concord), New Hampshire (Nashua), and— over the next ten years—on the West Coast (Visalia, California), and Puerto Rico (Ponce) and Italy (Milan). Electric power costs were also very high in North Adams; this forced the company to move such electrical power-guzzling capacitor families as Aluminum Electrolytics to southern regions, such as those served by the Tennessee Valley Authority.

II: SPRAGUE ELECTRIC JOINS THE SOLID-STATE REVOLUTION

Capacitor Assembly (North Adams Historical Society)

Like the company he ran, Robert Sprague also began to branch out. In 1945 he became a Director of the Associated Industries of Massachusetts, and was chosen President in 1951. Years later, in discussing this expansion in his personal horizon with his grandson, David, he commented, "It is my nature that, when I join something I either participate—I hope constructively—or get out."[13]

Since the late 1920s, all three of electrical inventor Frank Julian Sprague's sons, Robert Chapman Sprague (founder), Julian King Sprague (sales), and Frank Desmond Sprague (plant engineer/mechanical engineer) had worked for the Sprague Specialties Company.

Robert and Julian were key figures in the growth of Sprague Electric. Although their skills complemented each other, their very different educational backgrounds and personalities led to an increasingly turbulent relationship as the company grew and matured. Robert Sprague was a combination of engineer and businessman, but had ambitions well beyond his company and even beyond the electronics industry. Although lacking Robert's broad technical skills, Julian was no less brilliant and saw himself as much more than the consummate salesman he became. A proud man, with time he began to chafe as, organizationally, he was always forced to play second fiddle to his older brother.

Julian King Sprague was born in New York City on 14 June 1903, the second son of Frank J. Sprague and his second wife, Harriet Chapman Jones. He went to the Hotchkiss school and then Yale University, which he never finished. After several different jobs, he joined Sprague Specialties in 1926 as production manager. Several field sales jobs followed, and he rose rapidly within the sales organization leading to his 1945 election as Vice President for Sales. One of Julian Sprague's most important innovations was the creation of an applications, or field engineering, department to complement and support the field sales organization as the Sprague Electric portfolio of products continued to broaden and diversify. He also became deeply involved in many industry and government activities, and in 1954 was appointed chairman of the Advisory Board on Electronic Parts for the Department of Defense.

II: SPRAGUE ELECTRIC JOINS THE SOLID-STATE REVOLUTION

When Robert Sprague went to Washington in 1953 to accept nomination as Undersecretary of the Air Force (later withdrawn), the Board of Directors elected Julian president of Sprague Electric (although not CEO), a position he held until his untimely death in 1960. Even as illness began to overwhelm him in the latter 1950s, Julian continued his frantic work schedule, including a 1959 visit to the USSR to evaluate Soviet manufacturing techniques. Although a lifelong smoker and diagnosed with lung cancer in April 1960, in May he chaired a nine-person team charged by the Department of Defense with seeking cooperative ways of sharing research with France and England in electronic parts and materials.

In July 1960, Julian Sprague entered Massachusetts General Hospital to undergo radical surgery and radiation treatment, but to no avail. He died on 27 September at his Big Bend ranch in Texas.

Robert's two sons, Bob, Jr. and John also both played important roles in the history of the Sprague Electric Company. Robert Chapman Sprague, Jr. (Bob, Jr.) was born in Brookline, Massachusetts on 22 December 1922. He attended Middlesex School and one year of Williams College before Pearl Harbor ended his college education. Already an experienced pilot, during most of World War II he served as an FAA Flight Instructor for the Army Air Corps. Bob, Jr. joined the company in 1946 in the human resources department. He had strong mechanical skills—as a hobby, in later life, he collected and restored Stanley

Steamer antique cars—and what he knew best was how to fly airplanes. He rose through the human resource discipline to become a senior vice president. In 1951, he started Sprague Flight Operations, a small but active corporate department used primarily to ferry management, sales, technical, and support personnel to branch plants along the East Coast.

Two years later he joined the Sprague board of directors, and in 1958 he completed his renewed formal education when he graduated from MIT as a Sloan Fellow. Promoted to Senior Vice President in 1960, he added responsibilities in purchasing and facilities management and in 1964 became the senior vice president of Corporate Services, a position he held until his retirement in 1980. Besides his corporate responsibilities he was active in electronic industry and local civic affairs, and was also a colonel in the Civil Air Patrol and Commander of the Northern Berkshire Squadron. Continuing his passion for flying after retirement, in the middle 1980s he built a high performance Christen Eagle biplane from a kit. He was flying this plane when he was killed in a crash on 10 April 1987 at the North Adams Harriman and West Airport.

Another key executive, Ernest L. Ward, joined the company in early 1946 as a vice president. Ward's background was investment banking, yet his Sprague responsibilities began in manufacturing.[14] Previously he had been located in Chicago as a partner of the investment banking firm F. S. Moseley. As manufacturing was being diversified to increasing numbers of branch plants throughout the United States and the world, what the Sprague company needed was a brilliant executive who could attract and motivate good people, and effectively evaluate and improve their performance. Ward moved rapidly up the management ladder, first to executive vice president in 1952 and then president in 1960 after Julian Sprague's untimely death.

As 1946 ended, the company had nearly completed its peacetime conversion and once more was pursuing consumer markets such as AM and FM radio, automotive radios, phonograph console systems, and television—the next great market opportunity. As the Cold War intensified, Sprague Electric used certain basic strategies: pursue civilian and military markets, offer performance and reliability superior to the competition, continue component miniaturization, broaden the product portfolio, and seek geographic plant diversification. Ten years after the end of World War II, the four manufacturing locations (Beaver Street, Brown Street, and Marshall Street in North Adams, plus Barre, VT) had grown to thirteen worldwide and employment had more than doubled.

New products fueled much of the growth, many of which had been developed internally during the World War II: "coupling" capacitors to enable telephone service in rural areas over electric power lines, pulse-forming networks for radar equipment, noise-suppression filters for both civilian and military equipment, miniature molded tubular paper capacitors (Duracap®), and a line of miniature metal-clad hermetically-sealed capacitors for military equipment such as proximity fuzes and guided missiles. In addition, the Nashua plant (acquired in 1948) was awarded a large government contract for radio proximity VT fuze printed-circuit amplifiers (complex assemblies of resistors and capacitors deposited on ceramic substrates). This technology later developed into a major business in coupling and decoupling networks for the computer industry, especially IBM. Nashua was the last of the former textile buildings added to Sprague Electric's growing portfolio of worldwide manufacturing locations.

Not all the programs were successful. In 1948 the company introduced an innovative mercury hearing aid battery. This seemed a logical extension of Sprague's broad background in electrochemistry. However, the company was never able to make a commercial success of a series of battery initiatives, and eventually abandoned the effort in September 1951.

Growth also resulted from acquisitions, such as the 1948 purchase of Milwaukee-based Herlec Corporation which provided Sprague Electric the technology for disc ceramic, multi-layer vitreous enamel structures, and printed-circuit applications. These technologies led to Sprague's development of the multi-layer ceramic capacitor, a product family that today dominates the entire capacitor industry, although Sprague Electric was never able to capitalize on its first-to-market position.

Discrete capacitors constituted the core of Sprague Electric's business, a business it eventually led worldwide. However, as the company expanded its product base into other types of electronic devices and materials, there was one more ingredient about to enter the mix and revolutionize not just the electronics industry but the entire world.

Select Sprague Electric Company Components circa 1950 (North Adams Historical Society)

The Transistor

Passive electronic components (such as capacitors, resistors, and inductors), although crucial in any electronic circuit, cannot amplify electrical signals. Until the late 1940s, amplification could be accomplished only by using vacuum tubes. Even in miniature form, vacuum tubes were large and consumed a great deal of electrical power. As circuitry became increasingly complex (such as in the early digital computers ENIAC and EDIAC that were developed toward the end of World War II), the rate of failure of the tubes made for notoriously unreliable devices. The holy grail in electronics was solid-state amplification, which had been pursued worldwide since the 1920s. Just prior to World War II, one of the greatest collections of technical talent was concentrated on this goal in Murray Hill, New Jersey, at the Bell Telephone Laboratories, at the time the research and development arm of AT&T.

During World War II, this work was temporarily discontinued in order to concentrate on the wartime communications business of AT&T. Wartime experience had clearly demonstrated the need for a solid-state amplifier, and extensive government-sponsored research and development on germanium (Ge) and silicon (Si)— the principle materials used in microwave radar detectors during World War II— laid the basis for the first operational solid-state amplifiers. In late 1945 Mervin J. Kelley at Bell Labs created the Solid State Research Laboratory with the mission of developing a practical solid-state amplifier using either Ge or Si. A little over two years later, on 23 December 1947, John Bardeen and Walter H. Brattain demonstrated such a device, a crude germanium point-contact transistor. One month later, William B. Shockley conceived the more practical junction transistor, whose first operational realization appeared in early 1950. For their discoveries the three men shared the 1956 Nobel Prize for Physics.[15]

At first these results caused only a modest stir, but as their importance began to sink in, corporations from around the world sought license agreements with Western Electric, the production arm of AT&T.

II: SPRAGUE ELECTRIC JOINS THE SOLID-STATE REVOLUTION

By the time of the third Western Electric "Transistor Symposium" in April 1952, there were twenty-six United States and fourteen non-U.S. attendees. Included in the list of licensees was Sprague Electric. For Sprague this was a monumental decision, and in the minds of some, a fatal mistake.

While Sprague Electric was trying to figure out how to enter the transistor business, it launched an internal research and development effort in North Adams which included hiring a number of engineers who were being brought into the United States by the U.S. Signal Corps. The most important and talented of these was Czech-born Dr. Kurt Lehovec, who was hired by Preston Robinson in mid-1952 to head the Sprague Electric semiconductor research effort. Since Bardeen, Brattain, and Shockley had announced their discoveries, a number of different approaches for making germanium or silicon transistors had been developed. Because of production difficulties with the original point-contact device, the junction transistor quickly became the preferred device structure, where the required dopants were introduced into a single crystal germanium (or silicon) die or wafer by alloying, high temperature diffusion, or a combination of both. Lehovec first developed an improved point-contact transistor and then started to work on junction devices, including the creation of multiple p-n junctions in crystalline germanium. This led to U.S. Patent #3,029,366, "Multiple Semiconductor Assembly" (filed 22 April 1959 and issued 10 April 1962), Sprague's first integrated circuit patent.

In September 1958, Jack Kilby of Texas Instruments demonstrated the first semiconductor integrated circuit. After the related patent issued in June 1964, Texas Instruments began patent interference proceedings against integrated circuit manufacturers around the world, including Sprague Electric. Although Kilby's circuit was not a producible device, it proved the concept and therefore the potential financial implications from licensing were huge. Panic followed throughout the industry, especially among the smaller manufacturers such as Sprague Electric, which had very limited semiconductor patent portfolios of their own to cross-license. At first even Lehovec's patent didn't seem to carry much weight because it described a device structure that was never going to be industry practice. At Lehovec's request, Robert Sprague's son, John Sprague, studied the patent in detail. As he did so, he became increasingly excited. Almost hidden in the description, claims, and illustrations, Lehovec had patented the basic technique used throughout the industry for electrically isolating the different elements within an integrated circuit. Initially Lehovec himself disagreed, but he and John Sprague soon reached consensus, and therefore Lehovec and the Sprague patent attorney were well armed for the interference hearing with Texas Instruments that followed.

In March 1966 the U.S. Patent Office ruled that Lehovec owned the basic patent covering p-n junction isolation in integrated circuits, far and away Sprague Electric's most important semiconductor patent. Although Kilby was awarded the 2000 Nobel Prize in

Physics for his 1958 invention, Kurt's work earned him recognition as a co-inventor of the integrated circuit[16].

Philco Corporation, until this time known primarily as a radio manufacturer, took a different direction, and in 1954 invented the surface-barrier transistor. In the surface-barrier transistor, electrochemical etching accurately defined an extremely thin base region, and doping metals were electrochemically deposited on this base. Along with improved next-generation devices such as the micro-alloy type and the micro-alloyed diffused base, the Philco transistors were the world's fastest switching transistors throughout the remainder of the 1950s, and were used in the first commercially-available solid-state computer, the Philco S-2000 Transac introduced in 1957. It was the Philco approach that Sprague initially decided to follow in its transistor venture.

It is unclear how Sprague Electric made this decision, and it appears that Lehovec had little if any input. Perhaps it was the device's speed advantage, although this would disappear in late 1959 with the Bell Labs invention of the epitaxial transistor[17]. Philco had for many years been Sprague's largest customer, and there were excellent relations at the management level. Also the Philco process involved a high degree of mechanization, which was required because each device die had to be processed on an individual basis.

The competing planar transistor process, (or mesa, from its visual similarity to the rock formations of the American Southwest) which evolved from the basic junction transistor with all its initial limitations, was nonetheless a much more efficient batch process. By 1960, oxide masking and junction protection, combined with epitaxy in a planar construction with a flat surface, would not only eliminate the limitations in the junction transistor process, but also allow deposited surface interconnections to form integrated circuits. Planar technology led by Fairchild Semiconductor Corporation (a subsidiary of Fairchild Camera and Instrument), came to dominate the semiconductor industry.

Nonetheless, Sprague continued to pour money into the electrochemical approach in its Concord, New Hampshire, plant, and by the mid-1960s had built a modest annualized $10 million semiconductor business with a forecast of doubling it over the next year or two. By 1973, electrochemical transistors had become only a small specialty business at Sprague. By then Sprague Electric had jumped full force into planar technology, which, with many variations, is the dominant technology still used in the more than $300 billion worldwide integrated circuit market. Appendix 2 gives a more detailed time line of the development of the planar process.

"R. C." Goes to Washington

On 12 January 1953 Robert Sprague was nominated to be the Undersecretary of the Air Force. He resigned as president (Julian K. Sprague was elected to replace him as President and Bob Sprague, Jr. to replace him as a director). Four weeks later, the assignment had to be withdrawn because of Robert Sprague's unwillingness to sell his substantial stock interest in the Sprague Electric Company which—although not required by law—the Administration felt necessary in view of action by the Senate Armed Services Committee on the appointments of Charles E. Wilson, Robert Keyes, and two Service Secretaries who also had substantial stock interests. On 24 March 1953, Sprague rejoined the Company as Chairman of the Board of Directors.

In David Sprague's 1985 interview, Robert Sprague provided additional colorful insight into his aborted appointment. "Your grandmother was violently opposed, arguing that she didn't like Washington and wanted nothing to do with politics which, in her opinion, was a 'dirty business.' She even went so far as to say that if I decided to go to Washington it would be without her. Not really believing she would carry out this threat, I accepted the nomination, subject to the results of scheduled follow-up meetings with Secretary of the Air Force nominee, Harold Talbott (who would be my boss), and President-Elect Eisenhower. One of the key questions was whether or not I would have to sell my Sprague Electric stock to eliminate any conflict of interest, as had other nominees such as Talbott himself (a director of Chrysler), and Secretary of Defense Charles Wilson (former General Motors President). Since I had resigned all my Sprague Electric positions and Sprague Electric did not supply assembled end equipment either for civilian or military use, I didn't see a problem and Eisenhower agreed. However the Senate Armed Services Committee had other ideas and when I was asked directly by Talbott and Wilson if I would agree to sell my stock and I declined, the nomination was withdrawn, and in early February 1953 I returned to the Berkshires, at least for the moment."

By coincidence, Robert's son, John Sprague, was also in Washington during part of the time his father was going through the confirmation process. John Sprague was First Lieutenant and Gunnery Officer on the *USS Kleinsmith*, a destroyer escort converted to handle underwater demolition teams, and was attending the Navy's 20- and 40-mm gunnery school in the District of Columbia. John Sprague could tell that his father had been bitten by the Washington bug, both because Robert Sprague was certain he could make a real contribution, and because it would be an entirely new adventure. Years later Robert Sprague told John, somewhat sheepishly, that he really got bored just running Sprague Electric. He wanted new challenges, and new peaks to climb. Robert Sprague was deeply disappointed when his nomination fell apart. The disappointment was short-lived, however, as only seven months later he became even more deeply involved in the issues of continental defense and nuclear deterrence between the United States and Russia. When asked why he refused to sell his Sprague Electric shares, Robert Sprague explained that this might have allowed control of the company to pass to some outside entity and thus adversely affect the North Adams economy and the thousands of its citizens employed by Sprague Electric. He said he was unwilling to let this happen.

Julian Sprague had been extremely excited about the opportunity to lead the company as president. He was just as disappointed when, only a few months later, his brother was back again as chairman and still in charge. There were inside reports of violent arguments between the two, and at least one public record that Julian opposed involvement in semiconductors, while Robert saw this as the next great leap forward for Sprague Electric.

The period during the beginning of the Cold War, especially after the USSR exploded its first nuclear bomb in 1949, was a precarious one for the world. The tension intensified when *Sputnik*, the first man-made satellite, was put into orbit by the Soviets in October 1957. People who could afford it were beginning to include bomb shelters in their new homes. New museums such as the Clark Art Institute were located in places like Williamstown, Massachusetts, rather than in New York City or Boston, to hopefully escape any potential nuclear blast zone; and a new term, mega deaths (millions of casualties) began to appear in secret Government-sponsored reports that analyzed the results of different war-gaming scenarios. The Sprague *Log* 9 February 1951 issue included a sobering insert, "Survival Under Atomic Attack."

In September 1953, former Massachusetts Governor Leverett Saltonstall (who was now a United States senator and chair of the Preparedness Sub-Committee of the Senate Armed Services Committee) approached Robert Sprague with a request to chair a committee with the task of making an in-depth study and report of the Russian nuclear first-strike capability against the United States using intercontinental bombers. Because of the sensitivity of the information he received, Robert Sprague ended up being the only member of this "committee," although he did have staff support and full-time use of former President Roosevelt's old "Air Force One," a DC-4. Although he never resigned as CEO of Sprague Electric, the next four years of his life were consumed with this assignment. More detail concerning this period can be found in Appendix 3.

With submission of Robert Sprague's report, and the nearly simultaneous appointment of James Killian as Eisenhower's Science Advisor in 1957, Robert Sprague's formal involvement in continental defense was over except for his position as a trustee of the MITRE Corporation, and he returned to Sprague Electric as full-time chairman of the board, treasurer, and chief executive officer.

The Doldrums: 1952 to 1958

Despite flat revenues, 1952 to 1958 was a volatile period of transition for Sprague Electric. The costly and dangerous Cold War and arms race with the USSR continued. Both nations enlarged their nuclear arsenals, headed into space, and expanded their conventional weaponry. Although Sprague Electric's military sales gradually decreased from 1952 to 1958, this decline was almost exactly compensated for by expansion in television. At the same time, the transistor and the integrated circuit changed the design and component requirements for electronic systems. This seemed a perfect environment for Sprague Electric with its reputation for superior performance and high reliability.

More worrisome were the expenses between 1953 and 1958 of adding facilities around the world in anticipation of the next growth period, and skyrocketing overhead costs, especially in semiconductor research, development, and engineering. In 1952, research and development expenses totaled approximately $1 million, or about two and a half percent of sales. By 1966, they were just under $10 million and more than six and a half percent. In the passive component industry, such expenses tend to run around three percent of sales, while in integrated circuits the number is double or triple this number, and sometimes even higher. Sprague Electric needed a major success in semiconductors in order to afford the overhead structure it was creating.

Sprague Electric's most important new product of the 1950s was the 150D solid tantalum (TANTALEX®) capacitor. The 150D was a hermetically-sealed axial device whose initial applications were for transistorized circuits in airborne electronics, including guided missiles, computers, navigation and radar equipment. Its small size, high volumetric efficiency, shock resistance, temperature stability and use of a solid—rather than liquid—electrolyte made it truly revolutionary. It quickly found use across the whole spectrum of transistorized circuitry, with industrial and military computers particularly important markets. The 150D was a key element in IBM's 360 family of computers, and one of the reasons why IBM became Sprague Electric's largest customer.

Although Bell Labs is often credited with inventing the tantalum capacitor, in fact work on it began at Sprague in the early 1950s led by Preston Robinson. His U.S. Patent #3,066,247, "Electrolytic Device", (filed 25 August 1954 and issued 27 November 1962 after bitter litigation with Western Electric) was recognized as the controlling solid tantalum patent. The extreme miniaturization of the device resulted from the high surface area of the sintered tantalum anode and the high dielectric constant (25 versus 9 for alumina) of the tantalum pentoxide (Ta_2O_5) dielectric formed on the pellet. The solid MnO_2 electrolyte, created by thermal decomposition of manganese nitrate ($Mn(NO_3)_2$), accounted for the capacitor's extreme stability. From conception, Sprague Electric enjoyed, or shared with Kemet, the leading position in worldwide tantalum capacitor market share, a market that was $1 billion in 1987. For comparison, the multilayer ceramic capacitor, emerging as the dominant capacitor family, was $3 billion, and the worldwide aluminum capacitor market was $2 billion.

Multilayer constructions (to increase the area and therefore the capacitance) were first introduced in the

1930s using stacked layers of natural mica. Because mica-based constructions had a low dielectric constant, relatively thick layers, and required an expensive process, mica was eventually replaced by synthetic dielectrics such as vitreous enamel, then by titania ($TiO2$), and finally by barium titanate ($BaTiO3$) formulations, whose dielectric constants can reach the thousands.

As with solid tantalums, Sprague Electric was an early leader in multilayer constructions. In 1958, based on work by Jack Fabricius and George Olsen using modified DuPont equipment originally designed to spray vitreous enamel layers, Sprague Electric introduced the first production multi-layer ceramic capacitor, named the Type 31 C MONOLYTHIC® capacitor. Its solid construction and high volumetric efficiency made it especially useful for semiconductor circuitry. Multi-layer ceramic capacitors enjoyed several advantages over solid tantalums: much cheaper materials, greater volumetric efficiency, and the ability to use a very high degree of mechanization in manufacturing. Although there are applications for all the different dielectric systems, multi-layer ceramic capacitors came to dominate the market, led by Japanese firms. Despite early advances, Sprague Electric never succeeded in being a factor in the multi-layer ceramic capacitors segment of the industry.

Break-Out: 1959 to 1966

As Sprague Electric became an increasingly important factor on the world stage, in 1959 Robert Sprague began to expand the board of directors by adding outside members who had broad experience and name recognition. The first of these was Frederick R. Lack, an industry veteran who had retired after forty-seven years in the Bell System. Until his retirement from Sprague Electric in 1974, he was a close confidant of Robert Sprague and an early mentor and friend of John Sprague after he joined the company.

1959 also marked creation of a more formalized system of long-range planning with the establishment of the Fourth Decade Committee to plot a course for Sprague profitably to reach a minimum sales level of $100,000,000 by 1967, which would parallel the Electronics Industry Association growth estimate of eight percent for the industry. Robert Sprague was the committee's driving force, and he was joined by Julian Sprague, Ernie Ward, Neal Welch, Bill Nolan, Bill Lazier, R.C. Sprague, Jr. and Preston Robinson. Initially, no one on the committee knew anything about semiconductor technology.

Nonetheless, Sprague Electric was already involved in early work in semiconductors, and Sprague engineers played an important role in the integrated circuit's early development, beginning with Lehovec's PN Junction Isolation patent. Despite this pioneering effort, it would take Sprague Electric another thirteen years of struggle to create a viable integrated circuit strategy and business. By the time it had—except for tantalum capacitors—Sprague Electric had lost its world dominance in all the other dielectric families.

II: SPRAGUE ELECTRIC JOINS THE SOLID-STATE REVOLUTION

In 1959, Sprague Electric revenues surged with the end of a 1958 U.S. economic recession, strong growth in entertainment electronics (especially television), robust military purchases, and new products such as the solid tantalum capacitor. By 1966, the single most successful year in the company's history, profits reached $8.7 million on revenues of $141.5 million, and the company employed 12,500. The period of 1959 to 1966 was strong for the entire domestic electronic industry, which reached nearly $20 billion in revenues by 1966, driven by industrial computers, data-processing, color TV, and weapons systems for the Vietnam War. Although Sprague Electric was still feeling its way in semiconductors, it continued to gain market share in all the capacitor families, with the exception of the still-infant multi-layer ceramics market.

With the deaths of Julian Sprague and of directors Frank Bond and Harry Robbins, more outsiders joined the board of directors, including Joseph A. Erickson (former President of the Federal Reserve Bank of Boston) and two prestigious MIT Professors, Dr. Jerrold R. Zacharias (inventor of the atomic clock) and Dr. Jerome B. Wiesner. Wiesner resigned in January 1961, when he was appointed Science Advisor to President John F. Kennedy, but rejoined the board in 1964.

As Sprague Electric's product portfolio continued to broaden, centralized control became less effective, and the company began to create a series of operating divisions, each with their own manufacturing, engineering, and product marketing. The first of these, in 1958, was the Special Products Division, which manufactured magnetic components and discrete component assemblies, followed by the Transistor Division in 1960. Toward the end of 1960, the Resistor Division was created, and in 1963 Fred Scarborough formed the Filter Division. Nonetheless, overall management of the capacitor operations remained centralized for many more years, and these new divisions were not true profit and loss centers. All products were still sold through the powerful national sales organization (under vice president Carroll Killen), and pricing control was still centralized under Neal Welch, senior vice president for marketing and sales, and his inside sales group. However, this structure began to change in 1964 when the newly-formed Semiconductor Division took control of its own pricing.

As the semiconductor effort began to absorb more of the company's resources, cultural differences became more evident between "old Sprague" (primarily passive components, with centralized control by older experienced managers, industry leadership and image, and the internal cash generator) and "new Sprague" (primarily semiconductors, with decentralized profit centers, much younger and often less-experienced managers, initially little industry image, and a consumer of the cash generated by passives). This problem of the two Sprague Electrics was never completely resolved during the remaining life of the company. This was the situation that John Louis Sprague, Robert Sprague's younger son, stepped into in 1959 after receiving his PhD in Chemistry from Stanford University.

John L. Sprague was born in Quincy, Massachusetts on 5 April 1930, but grew up in Williamstown where he attended local schools, and then went away to Middlesex School in Concord, Massachusetts. John Sprague entered Princeton University in 1948 and graduated in 1952 with an AB in Chemistry as well as a NROTC commission in the U.S. Navy. He spent the next three years on active duty in the Navy where his first ship was the *U.S.S. Kleinsmith*. In the fall of 1953, he enrolled in the U. S. Navy's Officers Electronics Material School at Treasure Island in San Francisco Bay.

His last duty posting was aboard the Navy's first (and last) command cruiser, the *U.S.S. Northampton* (ECLC-1/CLC-1), as Assistant Electronics Officer and Electronics Division Officer. Loaded with electronics, the *Northampton* served as flagship for the Sixth Fleet in the Mediterranean from the fall of 1954 until the spring of 1955.

In the fall of 1955, Sprague entered Stanford University as a Ph.D. candidate in Physical Chemistry. A professor at Stanford, Claudio Alverez-Tostado, was studying silicon chemistry, and became Sprague's thesis advisor. Sprague happily settled into an unused dark room in the basement of one of the chemistry buildings, and began making crude germanium and silicon diodes and transistors. During his research Sprague received advice and some basic electronic equipment from Sprague Electric where researcher, and former Luftwaffe pilot, Rainer Zuleeg, was particularly helpful. Sprague was able to complete all his requirements by the spring of 1959. His doctoral thesis, "Studies on the Nature of Metal to Semiconductor Alloy Junctions," was particularly timely. Job interviews with West Coast start-ups, including Fairchild and Shockley Labs, went well until they learned that he was of the same Sprague family which owned the Sprague Electric Company. Subsequent visits to those start-ups' factories were summarily cancelled. Sprague received job offers from the Hughes Aircraft Thin Film Laboratory in Culver City, California, and the GE Labs in Schenectady, New York. From a research standpoint, however, the most exciting offer was the opportunity to work with Kurt Lehovec in the Sprague research labs.

In weighing the decision to work at Sprague Electric, John Sprague was concerned that the same problems that had occurred between his father and Julian might someday also arise between his older brother Bob and him (they did). John Sprague also worried that— no matter what he accomplished—in many people's eyes he would always be just the son of the founder. However, Sprague was determined to work in semiconductors, and the opportunity to work with Kurt trumped all such concerns.

With their two young sons, John and Bill, Sprague and his wife moved back to the Northeast, and John Sprague began work in April 1959. Except for summer jobs held by two of his children, John Sprague was the last member of the Sprague family to work for Sprague Electric.

John Sprague had earlier worked for the company during the summer of 1949 as a technician making electrical measurements on different types of capacitor dielectrics at 87 Marshall Street. His supervisor in that summer job was Dr. Taylor, who had been tortured as a Japanese prisoner of war during World War II, and as a result had to drink beakers of dilute hydrochloric acid before eating to replace the stomach acid which he could no longer generate. Each noon, Sprague joined the thousand or so employees that poured out of the main gate and headed downtown for lunch, a snack, to shop, or to just plain relax. Main Street was filled with shops, diners and restaurants, bars, several hotels and a movie theatre. The east end was dominated by church spires, and there seemed to be people everywhere, talking, laughing, and enjoying life. In 1949, North Adams was a wonderfully upbeat and exciting place to wander through and explore. His co-workers were friendly, although as the boss' son, he was treated as somewhat of a curiosity.

When Sprague returned to the labs in 1959 as a researcher, he had much more demanding responsibilities. Following studies of silicon surface states with Lehovec, his most rewarding project at Sprague Electric was in 1961 when he headed a team that developed Sprague's first planar transistor, and brought the company into the world of modern semiconductor technology. The team included three German World War II veterans, Dr. Carl Busen,

Dr. Hans Scheer, and Dr. Rudolph Dreiner, along with Joseph Lindmayer, a Hungarian refugee and brilliant designer[18], and Dr. James Casey, a former associate professor from the University of Rhode Island. The resulting SEPT® (silicon epitaxial planar transistor) was available for sampling within a year, and a family of companion integrated circuits or UNICIRCUITS® shortly thereafter. The success of these developments was remarkable, considering that Sprague Electric's planar work began from scratch only a year after the process became an industry standard (see Appendix 2).

In the early 1960s, the chairman of John Sprague's Stanford orals committee, Prof. Eric Hutchinson, spent a productive summer in the North Adams labs during which he made two key recommendations: investigate semiconductor doping using ions, and hire one of his former students, Dr. Kenneth Manchester (who was then at Shell Oil) to lead the investigations. Manchester, working with Dr. John MacDougall, filed key patents in 1964 on ion implantation, a method for introducing impurities into semiconductors that later dominated integrated circuit processing technology.

John Sprague's hands-on research activities lasted only a few years. Although he tried to maintain some personal research, by 1965 he was directing all of Sprague's centralized research and development initiatives. So that he could concentrate on semiconductors, John Sprague appointed Dr. Walter J. Bernard—who had been heading the Electrochemical Research

Division at Sprague Electric—as his assistant to oversee the passives research and development. This was a key position because passive components were generating all the cash and supporting the company as it moved down the semiconductor road.

Research, Development and Engineering

Robert C. Sprague at 1962 Dedication of Research Center (Author's collection)

As the 1960s began, research and development investment continued to surge, particularly in semiconductors. In October 1962, a new corporate research center was dedicated on Marshall Street across from headquarters. Staffed with scientists from all of the basic chemistry and physics disciplines, its mission was to investigate the physical and chemical properties of the materials used in manufacturing active and passive components, and to create pilot facilities for pre-production of the resultant products. Dr. Fred Fowkes (from Shell Oil) was hired as director of research, and a year later Dr. F. Lincoln ("Linc") Vogel (from RCA) as associate director.

In the semiconductor part of the center, two fully-equipped pilot lines were created. The first was for thin film ceramic-based hybrid circuits (CERACIRCUITS®), which were complex circuits with limited volumes, or where full integration was not yet possible. The second was for development of silicon-based integrated circuits (the previously mentioned UNICIRCUITS®). Initially, as had been done with the SEPT® planar transistor, products developed from both lines were to be transferred to Concord, New Hampshire, for production.

Norton Cushman transferred from the Special Components Division to head the planar pilot facility in the research center, and in 1964 he made a key hire when Dr. Robert S. Pepper, from University of California at Berkeley, joined the company to accelerate this effort. Besides assuming responsibility for completing the pilot line, Pepper began to assemble a linear circuit design team that would ultimately be one of the key elements in the Sprague semiconductor turn-around that began in the late 1960s. Besides being an outstanding scientist and technician, Pepper also had the ability to hire and retain top-notch talent. For years he competitively, and very successfully, raced hard track motorcycles, and on more than one Monday morning arrived for work battered and bruised, especially after one memorable weekend when a missed turn had landed him in a rock quarry. A social visit to his home usually found the living or dining rooms littered with the parts of one of his dismantled racing machines. Years later Pepper told John Sprague that the best thing that happened to him at Sprague Electric was meeting his second wife, Star, who at the time was employed at the Concord, New Hampshire plant.

Even in such a remote location as northwestern Massachusetts, Sprague Electric was able to attract many competent engineers and scientists. Partly this was because of the beauty of the northern Berkshire mountains, but Sprague Electric itself offered a number of features which were attractive to top scientists. It had consulting relationships at major universities such as Stanford, Penn State, MIT, RPI, and the University of Connecticut. Continuing education was encouraged, and the company supported cooperative programs at places such as the University of New Hampshire (near the Concord and Nashua plants) and

North Adams State College (now the Massachusetts College of Liberal Arts). For a while there was a special Sprague Electric-Williams College program where Sprague scientists taught courses in chemistry and physics at Williams, and engineers could earn a master's degree, studying on a part-time basis. In a few cases, such as Dr. Galeb Mahar, who was a Sprague Fellow and later R & D director, the company financially supported education all the way through the Ph.D.

Before the new research center existed, corporate research, development and engineering was primarily located in a series of laboratories and offices on two floors accessed by the Building 1 entrance, which was on the right after entering the main gate of the Marshall Street complex. This same entrance offered access to the executive offices. When John Sprague joined the company in 1959, he had a small lab on the third floor of adjacent Building 4. (When the Marshall Street plant was later converted to the Massachusetts Museum of Contemporary Art, this space became the museum's "tall gallery.") There were other labs full of equipment and offices – part of Kurt Lehovec's blossoming Semiconductor Research Department. Soon John Sprague and his technician, Howie Marsh, were joined by Warren Berner and then Otto Wied. Occasionally John Sprague would stop off to see his brother, Bob, in Human Resources on the left just behind the gatehouse, or his father on the second floor of Building 1 in his corner office (which still exists) on "mahogany row." One flight directly above, Neal Welch and his customer service department served as the nerve center keeping track of everything that was going on both within and outside the company.

Where the Massachusetts Museum of Contemporary Art's entrance now opens into a large lobby, there was a pleasant company cafeteria. There was no separate dining room for management, and the cafeteria was a great place to eat and talk with friends and associates, or access the rumor mill to find out what was really going on in the company. The Marshall Street cafeteria was available to everyone in the company, but many production and office workers either brought their own lunch or went downtown. With time, the other North Adams plants (Beaver Street, Brown Street, and Union Street) also had their own cafeterias.

The vast Marshall Street complex contained numerous laboratories where new materials and devices were created, analytical spaces for testing the make-up of incoming materials (including, at times, competitors' products), reliability test facilities, and screen rooms for studying interference filter problems.

The largest spaces were the production facilities. Massive chemical machines were used first to etch (in order to increase surface area) and then form (to create the dielectric layer) large rolls of aluminum or tantalum capacitor foil. In vast rooms women sat side-by-side at complex rolling machines (at the time, all manufactured internally within the company) which created the capacitor sections for insertion into cans or to be molded to make the finished product.

Before moving to Wichita Falls, Texas, the ceramic division had its own space in Building 6 at the far western end of 87 Marshall Street, where the two Hoosic River branches converge. Here great mixers prepared the ceramic materials for firing in huge kilns, either as discs or as multi-layer ceramic capacitors.

Worcester

In 1964, Robert Sprague decided that Sprague Electric should make a run at being a major supplier of integrated circuits. Although he recognized the already excellent work underway in the Sprague research center, he launched a much larger effort involving a major new facility staffed by an experienced team from outside the company. Products from the research center could be transferred in later. After an exhaustive search, Worcester, Massachusetts was chosen over several other seemingly more logical locations such as Silicon Valley in California. Worcester offered excellent financial incentives, was much closer to North Adams, and was likely to be less prone to the frantic turnover problems of the nascent West Coast semiconductor industry.

II: SPRAGUE ELECTRIC JOINS THE SOLID-STATE REVOLUTION

Worcester Semiconductor Plant (Author's collection)

Completed in 1966 after an investment of $7 million, the 132,000 sq. ft. facility was intended to have sufficient capacity to make Sprague Electric a viable competitor with any of the existing integrated circuit suppliers. Because the market potential was huge and growing rapidly, Robert Sprague believed there would always be room for another competent supplier, especially an established component firm with Sprague Electric's reputation. Breaking with his long existing capacitor philosophy of always trying to be first to market, the route Sprague chose was to become a second-source provider of existing digital logic families – in other words to produce and market copies of other companies designs, usually under license. Chosen for early production were diode transistor logic (under license from Signetics), resistor transistor logic (from Fairchild), and transistor-transistor logic (Texas Instruments). There was also a proprietary M514 proximity fuze program that was an update of the original World War II program. To jump-start this endeavor, the company began interviewing industry veterans to take over management of the new facility.

Les Hogan came to North Adams to offer his services and those of key members of his Motorola semiconductor team. Although Robert Sprague wanted to meet such an experienced semiconductor executive, he was not comfortable with hiring a complete team away from his close personal friend and business associate Motorola CEO Bob Galvin. Two years later Hogan left Motorola to attempt to rescue a foundering Fairchild.

John D. Husher was hired from Westinghouse (where, unlike at Motorola, no such personal relationships existed) as the Worcester plant manager. Husher brought with him a complete operating, technical, and marketing team, numbering more than two dozen people. Husher saw no need for the North Adams research activities, and didn't like reporting to Ken Ishler (then vice president of the semiconductor division). Although Ishler would eventually be gone, so would Husher, who was even less pleased with the next organizational changes.

Not everything of note in Sprague's glory years leading up to 1966 happened in semiconductors. In 1960, Sprague Electric's image was greatly enhanced by award of a $1,263,355 contract from the Autonetics Division of North American Aviation to develop components, with 100 to 1000 times the reliability of existing devices, to be used in the Minuteman Missile guidance and control system.[19] The resulting HYREL® solid tantalum capacitor not only satisfied the military requirement but was a key element in the computer and data processing market.

In 1962, Sprague Electric became a Fortune 500 company, and on 21 November 1966 it was listed on the New York Stock Exchange after its stock had been sold for more than twenty years over the counter. Solid tantalum capacitor sales continued to soar, leading to further expansion in Sanford, Maine. Even so, it soon became evident that the multi-layer ceramic capacitor segment was where the major growth in capacitors was going to be. The companies which would eventually dominate the multi-layer ceramic capacitor industries—such as AVX in the U.S. and Murata in Japan—dedicated their resources to that product family while Sprague Electric was dedicating most of its resources to tantalum capacitors and to semiconductors. To boost Sprague Electric's lagging position in this increasingly important segment of the capacitor market, a new multi-layer ceramic capacitor plant was opened in San Antonio, Texas. Robert Sprague's son John believes that, if Sprague Electric had dedicated the same level of resources to multi-layer ceramic capacitors as it had to semiconductors, it could have been a leader in that market.

Even in capacitors there was a competitive trend beginning as increased imports of consumer equipment from Japan increased their share of the market. Radios and television were the largest end-market for aluminum electrolytic capacitors, and an important segment for many other component families. For suppliers such as Sprague, this was a double-edged sword. Not only were major U.S. manufacturing customers being replaced by Japanese firms, but given the chance, these potential new customers preferred to buy their components from local sources in Asia. In testimony before the House Subcommittee on the Impact of Imports and Exports on American Employment, Robert Sprague complained, to no avail, about an unfair market that allowed foreign corporations free access to the United States while maintaining exclusion of U.S. products from their markets.

Although 1966 was a record year, and there was a modestly upbeat forecast for 1967, clouds were beginning to form. With North Adams and Worcester semiconductors going in different directions, there was growing concern about the costs of the total semiconductor initiative, about where semiconductors were taking the company, and who was really leading it. In early 1967, after decades of intense effort, the International Union of Electrical Workers (IUE) succeeded in becoming the bargaining agent for the North Adams production workers. The office and technical workers remained under an independent union, at least until 1969 when they became AFTE Local #101 under the AFL-CIO American Federation of Technical Workers. By then, all non-salaried North Adams employees were finally represented by national unions. Nevertheless, excepting the Barre, Vermont and Visalia, California plants, the North Adams plants were the only locations ever unionized during the life of the company. Following six months of difficult bargaining, three-year contracts were successfully negotiated in mid-1967, but further severe labor trouble loomed.

Instead of growing as forecast, Sprague Electric revenues dropped from $142 million in 1966 to $127 million in 1967, because of an unexpectedly depressed year for the components industry, although modest profitability was maintained. In 1968, the company recorded its first financial deficit since 1932, ($2.8 million) on revenues of $133 million. After a modest recovery in 1969, disaster struck in 1970 and 1971 – largely because of the 1970 strike in North Adams – with combined losses of just under $15 million.

Even as revenues dropped, overhead costs continued to skyrocket, driven primarily by the Worcester plant. By the end of 1967, the company's integrated circuits investment had reached $25 million, Worcester employment exceeded nine hundred people, and the digital logic second-source strategy had yet to generate sufficient revenue.

In August 1967, John Sprague, as senior vice president for semiconductor operations, was given responsibility for "fixing semiconductors." His purview included the Concord, New Hampshire transistor division, the Worcester plant, and the North Adams semiconductor research and development efforts. There were other important corporate changes as tough insider Neal Welch was named executive vice president, and a new passive component manufacturing executive, William E. McLean, joined the company.

As John Sprague became involved in the details of the Worcester operation, he concluded that the second-source strategy was not going to work. In capacitors, Sprague Electric had always been a technology leader, usually first to market, and therefore enjoying the advantages of early learning curve costs and pricing. John Sprague was certain it was too late to achieve the same dominance in semiconductors, and that only a niche strategy, where the Sprague Semiconductor Group could be a leader, had any chance of success.

The 1968 financial loss served as a wake-up call, even as the economic recession eased and sales increased modestly. The quarterly dividend was reduced from $ 0.15 to $ 0.10, and was eliminated in the 3rd quarter of 1970. Executive Vice President Neal Welch moved quickly and forcefully to increase efficiency and reduce costs "using a well-known firm of industrial engineers (the WOFAC Corporation) to introduce more disciplined work practices."[20] Using classic time-study techniques, the Work-Factor Corporation supported by on-site "VeFAC Analysts" introduced "VeFAC Programming," throughout domestic operations between early 1968

II: SPRAGUE ELECTRIC JOINS THE SOLID-STATE REVOLUTION

and mid-1969. Although there were undoubtedly some related savings, the program was itself expensive and resented by almost everyone, including direct labor operators as well as overhead personnel. More draconian measures such as mandatory salary cuts and layoffs introduced in 1970 added to the discontent.

John Sprague went further, transferring Bob Pepper and his entire technical group to Worcester, and leaving only a small materials research team in North Adams under Ken Manchester (who would also move a few years later). Because John Sprague's background was mostly technical and he had very limited actual operating experience, he hired industry veteran A. Normand Provost, who had spent the bulk of his business career at Texas Instruments, as Worcester operations manager. Husher resigned taking with him his entire team, and Norm Provost and John Sprague moved quickly to fill the resulting vacancies from both within and outside the company. Worcester and Concord were reorganized as decentralized profit and loss centers, as was the chip-and-wire hybrid circuit business unit, now under the direction of John Seacord. Provost also moved quickly to transfer the majority of the expensive "back-end" (assembly and test) operations from Worcester, first to Mexico and then to South East Asia.

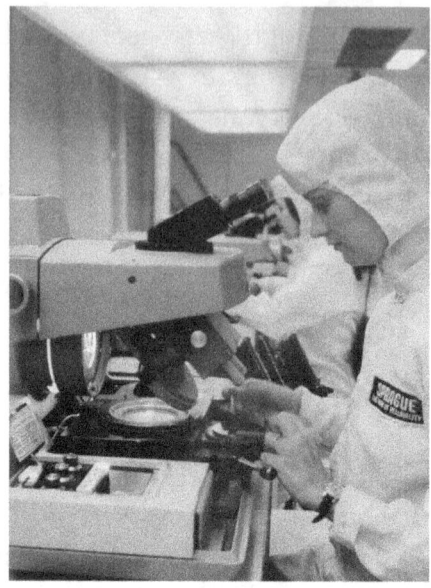

Worcester Wafer Fab (Author's collection)

Although the semiconductor division was still unprofitable, overall Sprague Electric finances continued to improve in 1969 with record sales of $147.1 million and modest profitability of $1.46 million. After several dismal years, capacity limitations again loomed, leading to the establishment of new passive component manufacturing plants in Tours, France; Renaix, Belgium; Taiwan; Scotland; and in Rhedyt, Germany.

In the mid-1960s, John Sprague and his wife had moved into a home they had built on Laurie Drive in Williamstown and where they had expected to live for the rest of their lives. However, to the great disappointment of John Sprague's entire family, this plan was about to change radically. John Sprague and his father Robert met there in John's study one 1968 fall evening for a quiet discussion about semiconductors. They had skied and played tennis together for years, and Robert had always supported John. Now John was going to challenge his father's strategy, in his father's most important new business, as being completely flawed.

John told his father that he had decided to move with his family to Holden, a Worcester suburb, after being told flatly by key semiconductor managers that no one would believe he was really committed to the division's success unless he did. John Sprague tried to explain that, without bankrupting the company, Worcester could only succeed as a niche supplier. Further, John planned gradually to phase out the different digital logic families and to bet the future of the division on the ability of Bob Pepper and his linear design team to develop unique linear integrated circuits for consumer and automotive applications. Already in the works were a television color demodulator for Zenith, a sound channel for Delco Radio (the Electronics Division of General Motors), a flash bar control circuit for the new Polaroid SX-70 camera, and a variety of other circuits.

Robert Sprague would have none of it, and as the exchange became increasingly heated and his face flushed red with anger, he called John a coward, someone who was either unable or unwilling to do in semiconductors what Robert had accomplished in capacitors. Deeply wounded, John quietly told his father that he would have to get someone else to run the semiconductor division. Realizing he had gone too far, and that John might be right, Robert Sprague quickly backed off, then reluctantly capitulated. John received conditional approval for his program, although things would never be quite the same again between them. An awkward parting handshake unexpectedly became a much warmer embrace, accompanied by a tear or two, and it wasn't long before it was the two of them against almost everyone else.

As the Sprague Semiconductor Group struggled for survival, two unexpected events helped brighten the otherwise gloomy outlook.

Mostek

Completely misjudging how it might react, the management of Dallas-based Texas Instruments informed their Metal Oxide Semiconductor group in the late 1960s that it was moving to Houston. MOS devices had become increasingly important because of their much lower power requirements compared to bipolar devices. The group rebelled, and led by L. J. Sevin and Lou Sharif contacted New Business Resources, a boutique venture capital group in Dallas whose principals, Richard Hanschen and Dr. Richard Petritz, had previously been Texas Instruments executives. New Business Resources next approached Sprague Electric. In mid-1969, Mostek was born, and for a modest investment of several million dollars Sprague found itself owning just below fifty percent of a fledgling MOS company with an extremely talented management group and staff.

It was a marriage made in heaven. Sprague had excess capacity in Worcester where Mostek could immediately begin processing silicon wafers. Mostek's discovery of Ken Manchester and John MacDougall's experimental single-wafer ion implantation reactor in the basement of the North Adams Research Lab turned out to be even more important. Initial experiments showed that implantation gave much tighter control of the gate surface potential of a MOS device, and therefore superior device performance over any competitor's. Convinced of its potential, Mostek bet the company on this new and untried technology. Wafers were processed in Worcester, then transported to North Adams for the gate implant, and finally back to Worcester for finishing. Led by the Semiconductor Division's new firebrand sales manager, Bob Duca[21], the Sprague sales organization helped out initially with customer contacts. While New Business Resources' Dick Petritz was Mostek's first president, it wasn't long before L. J. Sevin was running the show, and the company eventually consolidated its operations in Carrolton, Texas, while continuing to maintain a close relationship with Sprague (still their largest stockholder) both in technology and at the management level. In its brief ten-year life as an independent company, Mostek enjoyed phenomenal success, inventing the first single-chip handheld calculator, and becoming the early industry leader in the growing dynamic random-access memory market, until Japanese firms began to dominate that market in the late 1970s.

In addition to the eventual financial return, Sprague benefited in a number of other ways. Until the move to Texas, Mostek revenue helped cover Worcester overhead costs, while their success with ion implantation demonstrated that Sprague was a technology leader in semiconductors as well as in passives. More subtly, the personal and business relationships that developed between Sprague and Mostek benefited both organizations. In some ways, association with Mostek put Sprague Electric on the world semiconductor map.

Moon Wafer

Author with Moon wafer (Author's collection)

II: SPRAGUE ELECTRIC JOINS THE SOLID-STATE REVOLUTION

In mid-1969, Sprague Electric received an unusual request from NASA for a crash program related to the upcoming July launch of *Apollo 11*, the United States' first manned landing on the moon. That Sprague Electric should receive such a request was not unusual because the company had been deeply involved in the United States space program since its inception. However the nature of the request and the required timeline to accomplish the task were anything but usual.

Only a few weeks prior to launch, it had been decided that *Apollo 11* would leave a time capsule on the moon and in that capsule would be an artifact containing letters of congratulations from world leaders along with other information. The idea was to miniaturize all the text onto a single silicon wafer. Within Sprague Electric the only place such a task could be accomplished was in Worcester where an adaptation of the basic photolithographic process could place the messages on an oxidized one and one half inch diameter silicon wafer. Bob Pepper and his technical team were given the task of solving the problem and producing the wafers in an impossibly short period of time. The first messages were received on 3 and 4 July 1969 and the first complete wafer (plus spares) delivered to NASA on 6 July. As new messages came in, the whole process had to be redone. The final wafers, each containing messages from seventy-four heads of state, were picked up by a NASA messenger on 11 July. One of them was in place when Apollo 11 was launched on

16 July 16 1969 and it remains on the moon. There is a beautiful book by Tahir Rahman which describes the Mission and the silicon disc program.[22]

Buoyed by the successful completion of this seemingly impossible task, the Worcester group went back to work trying to save itself and, in the process, the Sprague Electric Company.

The 1970 Strike

Operational performance in Worcester continued to improve, and the 1969 Annual Report glowingly predicted Worcester profitability in the second half of 1970 (this didn't actually occur until late 1972). There were also several warnings: "At this time the outlook for 1970 is clouded by the uncertain general economic climate." The report continued, "Labor contracts with the three unions representing hourly employees in North Adams expired December 31, 1969 and although negotiations are continuing, at this time (March 5, 1970) no agreements have been reached."

Negotiations dragged on through January and into February. Walter Wood was the chief negotiator for the IUE, while the company was represented by John Winant, Bob Kelley, George Bateman, Bob Sprague, Jr. and—mostly behind the scenes—Robert Sprague. Although most of management couldn't imagine that a strike would actually happen after so many years of local prosperity and peace, the company prepared for the worst. Windows were boarded-up on Marshall Street, a detailed logistics plan was developed to deal with picket lines, and an emergency plan was created to keep some North Adams manufacturing going using salaried and non-unionized hourly employees. Sprague began to move as much manufacturing as possible out of North Adams, a process that accelerated during the strike itself. This drew a great deal of local criticism, but there was no choice. Sprague Electric had negotiated delivery schedules with its customers. In some cases these were for products where Sprague was sole source, while others were for crucial military and government programs. Inability to meet these commitments for whatever reason could be catastrophic in the short term, and cause long-term loss of key customers because competitors would rush in to fill the void. This is exactly what happened to some of the capacitor lines during the strike, where Sprague lost more than one third of its market share, a loss that took years to recover.

As the negotiating deadline of 1 March 1970 approached, positions on wages and benefits hardened. On Sunday 1 March, AFTE moved first, voting to strike by a margin of nearly two-to-one, and picket lines were set up around the Marshall Street complex which the IUE, and nearly all of the other hourly employees, refused to cross. Over the next ten weeks as management and salaried personnel continued to cross the lines and negotiations stalled, there were increasing incidents of violence; several cars were burned, and there was random heckling, along with pushing and shoving.

In a September, 2012 interview, former Mayor John Barret told John Sprague that most of the violence was instituted by outsiders the IUE had brought in to help incite the local strikers. "They quickly disappeared once it was settled."

Being located in Worcester rather than North Adams, John Sprague never crossed the picket line. However, his brother Bob did, twice a day, and Robert Sprague, who always went home for lunch, four times daily. For Robert Sprague, the daily passage across the line was devastating. It was as if his own family had turned against him, and relations between labor and management in North Adams would never be the same.

With little progress, in late April negotiations moved to Washington, DC under federal mediator J. Curtis Counts, and agreement was finally reached several weeks later. Wages were to increase six percent the first year of the three-year contract, and five percent each the second and third year. There were other improved benefits in such areas as vacation pay and hospitalization. On the company side, the much-maligned VeFAC program remained intact, and there was no requirement to bring back lost jobs. Although the company agreed to general amnesty for the striking workers, North Adams had become a much less attractive place to run a business.

The economic recession only made things worse. Compared to 1969, 1970 domestic electronic equipment sales decreased four percent, and component sales followed, dropping eight percent. In the meantime, component inventories soared to a more than a three months supply. (There is a vicious cycle that plagues the electronic components industry to this day, and which sophisticated computer systems have not been able to prevent: in hard times customers slash their inventories, leading to panic buying and multiple ordering when business improves. When inventories soar again, the cycle continues.) The 1970 annual report described, "Sprague Electric sales fell more sharply than the overall components industry reflecting loss of volume during the ten week strike as customers, fearing that the strike would spread to non-union plants, took their business elsewhere. (Besides layoffs) there was also an across the board salary cut on September 1, 1970 averaging eight and one half percent (one half of which was restored January 1, 1971) bringing the breakeven point to its lowest level in 5 years. Some industry improvement is expected in 1971."

Most of what has already been written about Sprague Electric relates to the company's labor relations in North Adams and to the 1970 strike, when the now nationally-affiliated unions representing the office workers (AFTE Local #101) and the production workers (IUE) decided to flex their muscles and show the hourly employees just how much they had gained by their new representations. Most of these writings strongly support the unions. However, on a practical level, as the U.S. economy softened, the strike couldn't have come at a worse time for management or for workers.

Wages and benefits in North Adams set the pattern for Sprague Electric operations across the rest of the country. In an increasingly competitive market threatened by foreign imports, Sprague could not afford what the unions were demanding which, in the case of the IUE, was compensation levels close to those at GE in nearby Pittsfield. From a product-cost basis, what really counted was not how Sprague Electric's hourly wages and benefits compared with GE, which operated in different businesses, but how they compared with its competitors which, according to EIA data, were on the high side in the U.S. and extremely high compared to Asian imports.

Ultimately both sides lost. Along with a depressed U.S. economy and electronics industry, the strike brought the company to its knees financially. Although hourly workers received some improvement in wages and benefits, the strike resulted in an accelerated movement of jobs out of North Adams, to southern facilities with new modern manufacturing facilities, lower wages and power costs, and to low-wage areas such as Puerto Rico, Malaysia, Taiwan, Hong Kong, and the Philippines. As a direct result of the strike, more than 2000 jobs left North Adams never to return. Over time many of these jobs would probably have eventually been lost anyway. However this would have happened at a much more gradual pace, and without leaving the same level of ill-will that tainted the post-strike relations between management and the North Adams labor unions.

II: SPRAGUE ELECTRIC JOINS THE SOLID-STATE REVOLUTION

If 1970 was bad, 1971 started out even worse, and component shipments began to strengthen only in the second half of the year. For the year, Sprague revenues totaled a dismal $117.9 million, and there was a staggering net loss of $8.10 million. At the operating level there were further layoffs, and the move toward decentralization accelerated as passive component product development was transferred into the operating divisions, and only a small centralized research activity remained in North Adams. While Worcester continued its strategic shift to linear integrated circuits, and Mostek surged with a broadening product base of calculator chips and dynamic random access memory chips, Sprague Electric's future in semiconductors was still very much in doubt.

Minutes of an early January 1971 meeting of the Executive Committee showed growing concern over the deteriorating financial condition of the company: "there are still too many expense personnel; if we are going to hang in Semiconductors for the future, it will be necessary for the balance of the business to support the activity; we need to immediately reduce weekly overall costs by $150 thousand, primarily by headcount reductions; what are we going to do about the aircraft?" And independent outside director Erickson posed the key question on so many minds, "So what are we going to do about Semiconductors (especially Worcester) where the prospects of making a profit seem poor?"

In a private 23 June 1971 memo to Robert Sprague (it was shared with the Executive Committee a week later), John Sprague argued the Worcester case: "we are making good progress but need more time; some of SEC's greatest talent is in Worcester; we are critical to the success of Mostek; long term the future of the Electronic Components Industry, and therefore Sprague Electric, lies in ICs." John Sprague had grown up in a home where failure was never an acceptable option, and he had no intention of changing this in Worcester. John Sprague also admits that he had become obsessed with saving the business.

Bruce Carlson countered with a much more conservative plan which the entire Executive Committee, except John Sprague, endorsed in late July: phase out the Worcester wafer fab; exit all semiconductor product lines except Concord's profitable SEPT® transistors and Worcester's linear integrated circuits; move the linear wafer lab to the North Adams Research Center and related assembly operations to Juarez, Mexico; and sell the Worcester plant.

After an impassioned debate by the full board of directors at its 23 November 1971 meeting, the board passed an ambiguous resolution to, "Continue, for the time being, without specification as to length of time or degree of intensity, in the IC business, subject to continuing BOD review, and upon condition that the Company take whatever might be reasonably necessary to maintain its competitive position in its profitable passive component lines." The only negative votes were cast by Arthur G. Connolly (who had completely lost confidence in the current semiconductor initiative and accused John Sprague of being solely responsible for destroying the company) and Bob Sprague, Jr. Any immediate plan to close Worcester was then dropped completely in December because of rapidly increasing integrated circuit orders. Worcester had bought more time as Sprague Electric's overall business once more began to surge. The semiconductor turn-around climaxed in the fourth quarter of 1972 when, for the first time, both Worcester and all semiconductor operations finally turned profitable.

Responding to heavy pressure from the outside board members, in early November 1971 founder, BOD Chairman, and CEO Robert C. Sprague resigned as head of the company and was elected honorary Chairman of the Board and chairman of a new executive committee (which also included Welch, Carlson, and Ward). Although no longer involved in day-to-day management, Robert Sprague was anything but marginalized, and the executive committee continued to be extremely active until the end of 1976, meeting at least monthly, and often weekly.

Reporting on Robert Sprague's resignation, a 10 November 1971 article in the *Berkshire Eagle* by Thomas Morton summarily dismissed the entire Sprague family: "(over 25 years) 'Junior's' (Robert C. Sprague, Jr.) main contribution has (only) been to direct the air arm of the corporate headquarters; John is credited with the Worcester fiasco; and "R. C." has become more infallible than the Pope." However, with the Sprague family owning nearly twenty percent of the outstanding Sprague Electric shares, it was a little early to write off their influence.

Inside Sprague Electric there was only one person who was not only acceptable to the board of directors, but also qualified to succeed Robert Sprague. That was Neal Welch who, having joined Sprague Specialties in 1932, knew the company inside and out. He also had excellent relations with such major customers as IBM, Western Electric, and Delco (known as the "Top Three" within Sprague Electric) who had to be wooed back following the strike. If Neal had a blind spot, it was lack of hands-on manufacturing experience. Woe to the plant manager who missed a delivery promise, no matter why, especially to the "Top Three."

John Sprague remembers an angry exchange he and Welch once had after a beleaguered Sprague factory had missed yet another IBM delivery (as it turned out caused by an unexpected radical change in the customer's delivery requirements): "Well they (the Sprague factory) did it again", to which John Sprague had replied, "Neal, *they* are us." "No, *they* are they!" As with many super salesmen, it sometimes seemed that Welch's allegiance to Sprague Electric's major customers was stronger than it was to the company he led.

There was widespread opinion that Neal would finally terminate the "Worcester experiment," but that didn't happen. Total company revenue and profitability began to rebound in 1972, then soared in 1973 and 1974, and Worcester continued to show steady improvement. As a result, the issue was dropped for the time being.

II: SPRAGUE ELECTRIC JOINS THE SOLID-STATE REVOLUTION

1973 Sprague Electric Board of Directors. Standing left to right: Robert Sprague, Jr., Art Connolly, John Sprague, Jerrold Zacharias, Joseph Erickson. Sitting left to right: Robert C. Sprague, Gus Kinzel, Bruce Carlson, Fred Lack, Jerome Wiesner, Ernie Ward, Neal Welch, Gordon Phelps. Missing: William Nolan (deceased January 1974), Bob Armitage (Author's collection)

In order to concentrate resources in the most profitable product segments, Worcester began a serious program of asset redeployment that included the Worcester-based chip-and-wire hybrid circuit operation, which was purchased by Hybrid Systems. Before the sale, this operation had enabled Worcester to pioneer a line of fully-integrated power interface circuits. This was the first of an increasingly complex family of power integrated circuits that became one of the major product families of Allegro Microsystems, the successor company of the Sprague Semiconductor Group.

Even as Sprague Electric rushed to meet current requirements, at a 26 February 1973 meeting of the executive committee, Market Research Director Len Lee cautioned that the current growth cycle would probably last only an unusually short two years, at least in part because of expanding imports of home entertainment electronics from Japan. It is unknown whether or not Lee also anticipated the OPEC oil embargo that ran from October 1973 until March 1974, triggering a world-wide recession.

Because of the success of the "Mostek Model" (growth by making minority investments in promising high technology start-ups), led by executive committee chairman Robert Sprague, who viewed this approach as a way to accelerate the company's growth, Sprague Electric made three similar investments between early 1973 and 1975: Boston-based *E R Corporation* (inexpensive high-fidelity audio speakers); *Micro-Bit* (electron-beam addressable memories); and *Princeton Material Sciences* (liquid crystal displays for consumer applications). All three had talented management and good ideas but, unlike Mostek, for various reasons all three failed and Sprague eventually had to write off the bulk of its related investments (roughly $3 million, plus the associated time spent by management and the R & D organization).

In 1974 Sprague Electric had record sales of $214.8 million, and nearly-record profits of $10.2 million. It was also a year of dramatically changing order patterns; surging first-half orders dropped dramatically in the second half of the year as the economy slowed and inventories soared. The full financial impact was felt in 1975 when sales dropped to only $161.9 million causing a loss of $10.3 million. Despite a quick return to profitability in 1976, 1975's dismal financial performance, coupled with the disasters of 1970 and 1971, led to 1976 being the last year Sprague Electric existed as an independent publically owned company.

Coincident with the deterioration in operational and financial performance, three additions to the board of directors infused new life, new objectivity, broad experience, new energy, and more controversy into a group that had become almost shell-shocked by the financial roller coaster of the previous years. In February 1974 Dr. David Ragone, Dean of Engineering at the University of Michigan, joined the Board of Directors, replacing Jerry Wiesner who had retired. Ragone and Robert Sprague were associates at MITRE Corporation[23]. That June, Dr. Robert Charpie, president of the Cabot Corporation and chairman of MITRE, came aboard, replacing retiring Fred Lack. Finally in May 1975, Sprague Electric elected its first woman director, MIT Associate Professor Dr. Margaret MacVicar. Very quickly these three new directors coalesced as a group, and their collective presence changed the direction of Sprague Electric, especially when Charpie began to view Sprague Electric as a possible acquisition candidate for Cabot. In November 1975, Bruce Carlson resigned as president over differences of opinion on his future role in the Company, leaving Welch to wrestle with the full management load while a search committee sought Carlson's replacement from candidates both inside and outside the Company.

By 1975 there were two very different views of Sprague Electric. In one, the company was a hidden jewel that was successfully reorganizing itself for renewed long term growth and stability. In the more prevalent view, generally favored by independent stockholders as well as many outside directors, Sprague Electric was a dying company, with a failed business model, too many marginal business units, too many facilities, too much overhead, and too many Spragues.

III: PROGRESS, THEN STRUGGLE TO SURVIVE, AND LIFE AFTER DEATH

Except for the early years, when ownership was concentrated within the Sprague family and a few family friends, Sprague Electric was a public corporation for most of its life. Public ownership, however, did not preclude the family influence. From the beginning right up to the mid-1970s Robert Sprague (regardless of titles, and even during a period of nearly full-time government service), was the chief executive officer and guiding light of the company. His brother Julian, as well as his two sons also played significant roles. This changed in the years following the strike, as the company struggled for survival. First there was an aborted acquisition (by Cabot Corporation), an actual acquisition (by General Cable/GK Technologies), a second acquisition (by Penn Central Corporation), a spin-off (as Sprague Technologies, Inc.), and finally the disappearance of the company as a single entity by the early 1990s. Multi-billionaire Carl H. Lindner, Jr. from Cincinnati, Ohio had the final say on the demise of the Sprague Electric, and was the Company's last Chief Executive Officer.

There were some very good times before things began once more to turn sour in the middle 1980s, and, although during much of this period Sprague was well supported by its new owners, the company had lost control of its destiny. Much like Arnold Print Works before it, Sprague Electric—buffeted by recessionary cycles, increasingly strong foreign competition, and strategic and tactical mistakes—eventually vanished. The transformation of the Sprague building at 87 Marshall Street from an industrial complex to a contemporary art museum is an example of what happened within much of the U. S. manufacturing sector in the early 21st century.

Although the three new directors did bring new life and vitality to the Sprague Board of Directors, Margaret MacVicar widened the chasm that had been developing between the directors and Sprague management. In mid-1975, following two visits several months apart, she circulated a report on the Worcester integrated circuit facility to all the outside directors as well as to Robert Sprague, Neal Welch, Bruce Carlson, and John Sprague. "There is no research and development; general plant-wide consciousness is near zero in regard to vision, materials development, new devices, broad strategy, etc.; where will new ideas come from, who is thinking about them, who has time, is this Division desired and/or viable?" (John Sprague's personal papers) Although seething inside, John Sprague replied courteously both in writing and in person. Little else was possible since, after a profitable 1974, his original 1975 sales forecast for the Semiconductor Group (Concord, Worcester, and North Adams Semiconductor Research and Development) had dropped from a profitable $17.3 million to $12.2 million, well below the breakeven point. After-tax profit for the division in 1974 was $634,000 but a negative $1,481,000 in 1975.

In early September, MacVicar wrote a second report on Wichita Falls (Sprague Electric's main ceramic capacitor plant in San Antonio, Texas, which was now also John Sprague's responsibility), that was only slightly more complimentary. "Turnover problems are probably the fault of the plant and not the town; the management talent is the most uneven in regards to quality that I've seen in my limited Sprague plant travels, and the youngest; (still) the plant seems salvageable in the moderate term and not the hopeless situation I had been led to expect." (from John Sprague's personal papers)

Both her evaluations later proved mistaken. Wichita Falls was never successful and eventually was closed, as first AVX and Kemet—then later several Japanese firms—came to dominate the multi-layer ceramic capacitor industry. By contrast, what was originally the Sprague Semiconductor Group in Worcester became a success story in the semiconductor industry, but under very different ownership.

Cabot Corporation

In the 1975 Sprague Electric annual report (published in the spring of 1976), Chairman and President Neal Welch reported, "On March 9, 1976 Sprague Electric and the Cabot Corporation agreed in principal to combine the two companies with Cabot issuing 0.633 of a share of its stock for each share of Sprague which will become a wholly owned subsidiary of Cabot. On March 3, Dr. Robert Charpie, president of Cabot Corporation, resigned as a director of Sprague Electric to avoid any possible conflict of interest between the two companies." Detailed due diligence was carried out by both companies over the next two months.

As part of the due diligence, John Sprague made visits to several different Cabot locations. At the time, Cabot had three major businesses, performance chemicals (it was the world's largest producer of carbon black, used primarily as a reinforcing agent for rubber in automobile tires), engineered products, and energy. John Sprague was impressed with what he saw and the people he met, and everyone seemed friendly, if somewhat reserved. Still, these were very lonely visits, as John Sprague could not see how Sprague Electric, the Sprague Semiconductor Group, or he himself fit into the Cabot plans.

The deal fell apart. The Sprague Electric Company's 1976 annual report noted that "On May 4 it was announced by officers of both Sprague Electric and Cabot that they have, by mutual consent, terminated the previously announced merger negotiations." In one view, it was because Robert Sprague concluded that Sprague Electric would not be run the way he felt it should be. Another possible reason could have been reluctance by the Cabot family at having so much Cabot stock in the hands of members of the Sprague family. There were no apparent hard feelings. Charpie rejoined the Sprague board on 30 June 1976, and served as one of Robert Sprague's closest confidants over the next five months.

Because business and financial results were improving rapidly, and it appeared that Sprague Electric would continue to operate as an independent company, starting in mid-1976 there was a flurry of correspondence, reports, and meetings of the Executive Committee (Welch, Ward, Ragone, Charpie, and Robert Sprague, chairman) and the Board of Directors to discuss the future of the Company and its leadership going forward. On 30 June 1976, the board of directors passed a resolution stating preference for an insider to fill the vacancy created by the departure of Bruce Carlson as president.

At an 18 October 1976 Executive Committee meeting, to which all the independent outside directors were invited and present, Robert Sprague proposed a new organizational structure to be in place from November 1976 until November 1979 which included Welch and himself in their current positions, John Sprague as president, and Bill McLean as executive vice president. Not pleased, the outside directors asked for more time to consider the proposal, and one—Art Connolly—argued that John Sprague and the Worcester operations had been solely responsible for the company's financial woes since John assumed responsibility for Semiconductors in 1967. At this same meeting, Robert P. Jensen, president of General Cable Corporation, was proposed as a new director.

Things deteriorated so badly that, at the end of October, the outside directors requested representation by their own independent counsel, to be paid for by the company. Robert Sprague and Neal Welch agreed, as long as the sole issue to be discussed was succession to the presidency. Meantime in Worcester, John Sprague followed the proceedings from a distance by means of periodic input from his father, including blind copies of some of the more contentious correspondence. Not only did the presidency now seem completely out of reach, but John Sprague also began to wonder if he would even have a job when all the smoke had cleared, or if he wanted to remain with Sprague Electric as the object of so much ill will. Worcester had become his comfort zone, while return to North Adams in some senior executive position was now increasingly unattractive.

General Cable/G K Technologies

The 12 November 1976 issue of the *North Adams Transcript* reported "Rumors fly at Sprague as trading in its stock remains suspended at the New York Stock Exchange." Five days later it lamented "The passing of Sprague Electric from the effective control of its founder, Robert C. Sprague, and members of the Sprague family, to new ownership marks a major transition in the local industrial scene."

The details of the transaction were described in the Sprague Electric 1976 10-K (the annual filing that all publicly-traded companies have to file with the Securities and Exchange Commission) "General Cable Corporation announced in an Offer to Purchase dated November 12, 1976, and extended to December 6, a tender offer for any and all shares of Common Stock of the company at a price of $19.50 per share (prior to the tender Sprague common shares had been selling at approximately $11/share). By 11 March 1977 General Cable had purchased 3,312,745 shares or 95.3% of all those outstanding. On 12 December 1976 six of the Sprague Electric independent outside directors (Charpie, Connolly, Kinzel, MacVicar, Ragone, and Zacharias) resigned and were replaced by Robert P. Jensen, Dennis G. Little, Donald R. Kampman, David C. Searls, and Larry G. Morris, all from General Cable, and Hugh H. van Zelm an investment banker who was also Florence Sprague's brother. Of the legacy outside directors, only Joseph A. Erickson remained, as his offer to resign was not accepted. The 10-K also reported a dramatic turn-around in Sprague Electric's 1976 financial performance with net sales of $199.6 million, the second highest in the company's fifty-year history, and net income of $6.8 million, compared to a loss of $10.3 million the prior year. General Cable's timing was exquisite.

Also at the 20 December board of directors meeting, nearly the same management structure as Robert Sprague had unsuccessfully been trying to sell to his original board was put in place for the new General Cable subsidiary, with Neal Welch as chairman & chief executive officer, John Sprague as president and chief operating officer, Bill McLean as executive vice president, and Gerry Tremblay as senior vice president. The exception was that Robert Sprague relinquished the last of his management responsibilities, although he was elected honorary chairman of the Sprague Board of Directors. A new era for Sprague Electric began, an era that proved to be unusually productive until some of the old problems resurfaced in the mid-1980s.

In the 1976 General Cable Annual Report there was a lead photo of a smiling Neal Welch and John Sprague, followed by comments on financial performance, new products, Mostek, overseas operations, and the Sprague Products Distributor subsidiary, which now accounted for roughly one third of domestic sales. Multi-layer capacitors were once more noted as the dominant future capacitor family and Worcester continued successfully to refine a strategy built around specific, mostly niche, areas of product leadership. These included linear consumer integrated circuits, high-voltage, high-power interface circuits for computer and printer applications, and Hall Cell integrated circuits for automotive applications. The recovery from 1975 was breathtaking. Compared to 1976, by 1980 sales had more than doubled to $453.0 million and operating income nearly six times to $77.6 million.

Often when such mergers or acquisitions occur, the acquirer immediately starts to impose its culture on the acquired company, consolidate similar functions to reduce costs, put members of its management team in key positions, and thereby often destroys what was best in the purchased entity. This did not happen with the General Cable acquisition, both because the two companies were of similar size so that neither dominated the other, and because they were different types of business. In 1977 Sprague Electric contributed thirty-seven percent of the consolidated sales and nearly forty-four percent of the operating income. By 1980, the equivalent numbers were thirty-eight percent and nearly fifty-two percent of what was now called GK Technologies, and which included another acquisition—Automation Industries. Robert Jensen had acquired a going concern that he expected to run its own show, at least as long as it continued to meet its commitments.

Robert Jensen was an excellent manager, very demanding, but also fair. He always carried a large briefcase which everyone assumed was filled with important papers. John Sprague learned otherwise at a Mostek Board Meeting when he and Jensen briefly had to share a motel room because of a mix-up in the reservations. Jensen popped open the briefcase which, it turned out, was filled with huge cigars, one of which he immediately lit. John Sprague was later able to escape to another room which had become available.

Overall relations between Neal Welch and Bob Jensen were amicable, and Welch worked very hard to keep them so. Inevitably, however, problems did arise. On one memorable occasion, a major customer called Jensen directly when one of its production lines was shut down by delinquent Sprague deliveries. Furious, Jensen exploded at Welch, and on the spot threatened to replace him with John Sprague. Jensen cooled down quickly upon learning that the problem had been caused primarily by an unexpected increase in the customer's delivery requirements, and that everything was being done to increase capacity. However, Welch was shattered by Jensen's reaction, especially because the acquisition of Sprague Electric was beginning to make Jensen look like a genius.

III: PROGRESS, THEN STRUGGLE TO SURVIVE, AND LIFE AFTER DEATH

The author and Neal Welch under Sprague Electric Clock Tower, circa 1977 (General Cable Annual Report)

Welch and John Sprague had worked well together for years, although the relationship remained a bit prickly. Welch was "old Sprague," conservative, oriented toward the passives side of the business, and towards keeping the customer happy whatever the cost. John Sprague was "new Sprague," much younger, semiconductor-trained, and more of a risk-taker. It was never clear whether Welch agreed with the semiconductor initiative. When John Sprague once asked Welch how he felt John Sprague was doing, Welch replied, "you are a good planner," a less than ringing endorsement. Nonetheless, in the late 1970s, as a team Neal Welch and John Sprague were able to produce outstanding results.

While Sprague management struggled to keep up with the resurgent customer requirements, the company began to reevaluate its relationship with Mostek. Although the subsidiary continued to perform admirably, there were danger signals, especially in the important DRAM (dynamic random access memory) business where Japanese semiconductor companies were beginning a successful assault on the U.S. industry. Concerned about Mostek's future, and needing cash to support Sprague's skyrocketing capital requirements, in 1978 Sprague Electric sold 600,000 Mostek shares for $9.8 million, reducing its ownership from thirty-four percent to twenty-one percent. The following year, Sprague Electric sold the remaining shares for $51.5 million.

In 1979, United Technologies purchased Mostek for $345 million, then mismanaged the acquisition, eventually selling what was left of the Company to Thomson Semiconductor for $71 million in 1985. In the meantime, key Mostek executives had gone on to other distinguished careers. L. J. Sevin joined Ben Rosen in 1981 to form Sevin Rosen Funds, Vin Prothro founded Dallas Semiconductor in 1984, and Bob Palmer joined Digital Equipment in 1985, eventually replacing founder Ken Olsen as CEO in 1992.

Although no longer directly involved in the business or industry, Robert Sprague received probably his most important accolade in April 1979 when he was again awarded the Electronic Industries Association's Medal of Honor, the only two-time recipient. The following year he seemed even more pleased when another of his goals was finally realized. At a 19 December 1980 special meeting of the Sprague Electric Board of Directors, John Sprague succeeded Neal Welch as CEO through a change in the by-laws which added this responsibility to those of the president. Neal Welch continued as chairman.

Early in 1981, Jensen announced in GK Technologies' 1980 Annual Report that, "discussions with Penn Central Corporation (PCC) have culminated in an offer by Penn Central to purchase for cash GK common at $50/share and convertible stock at $59.50 resulting in GK becoming a wholly owned subsidiary of PCC. We expect that after the merger the Company will continue under present management and organization as a subsidiary of PCC." The purchase price of $704 million was $175 million greater than the asset value of GK Technologies, and the difference (or goodwill) was loaded onto the cost structure of the GK business units, especially Sprague Electric. The offer was accepted by the GK Board on 23 February. Within Sprague Electric there was almost universal shock: Penn Central, the bankrupt railroad company? What did it know about electronic components?

Penn Central Corporation

The Penn Central Corporation resulted from the 1968 merger of the two largest U.S. railroads: the Pennsylvania and New York Central Railroads, both with roots back to the early 1800s. When Penn Central filed for bankruptcy in 1970, it was the largest corporate bankruptcy in America history. After years of litigation, in January 1981, Penn Central received cash compensation of $2.1 billion from the U.S. government for its rail properties, which had become part of federally-funded Conrail in 1976. Flush with cash and a $2.2 billion tax loss carryforward, what had once been a bankrupt company was now very rich, and it quickly began a major acquisition program. One of the biggest was the 1979 acquisition of Marathon Manufacturing, a manufacturer of offshore oil drilling platforms, and several other energy related businesses. Other parts of Penn Central included its Living and Leisure Group (Arvida, Great Southwest, and Six Flags Corporations) and the Diversified Industries Group. There seemed to be little or no strategic fit within such an eclectic mix.

Although seemingly a minor part of the corporate structure of a more than $3.0 billion highly-diversified corporation, Sprague Electric was prominently featured in the Penn Central 1981 Annual Report: "One of GK's primary businesses is conducted by Sprague Electric Company. Sprague is one of the world's leading manufacturers of capacitors, a vital component used in almost all types of electronic equipment sold around the world."

Although recessionary pressures in the 1981 economy had caused a decrease in Sprague's income, the Penn Central 1981 Annual Report predicted that "a recovery in the domestic economy during 1982 should result in improvement for all operations and the longer term forecast for electronic components continues bright." Nonetheless, the drop in earnings was the first since 1976, after five consecutive years of improved performance. Based on Sprague Electric's history of cyclical performance, this should have served as an early warning that the electronic component industry was probably approaching another of its recessionary cycles.

The top management of Penn Central changed as Al Martinelli, previously president and CEO of the Penn Central Energy Group, replaced Richard Dicker as president and CEO. There were two additions to the Board of Directors which would have a profound impact on the future direction of the company, Carl H. Lindner, Jr., a brilliant self-made Cincinnati billionaire and CEO of American Financial Corporation, and his right-hand man, Ron Walker. By 1983 Lindner had accumulated sufficient stock to take over control of Penn Central and the chairmanship of its board of directors.

Jensen moved on to Tiger International where he was elected chairman and CEO in 1985. Within Penn Central, Gene Swartz moved up to fill the vacancy left by Jensen as president of the Electronics, Defense and Telecommunications Group, thus becoming John Sprague's new boss for a short period of time. Senior Executive Vice President Herbert S. ("Pug") Winokur, Jr. followed as John Sprague's immediate superior, a relationship they both enjoyed for the brief period it lasted. As the result of all these changes—at a time when Sprague Electric could least afford it—it was imbedded in an initially supportive organization with good intentions but with minimal experience in the components industry and— in the case of Lindner and Walker—little desire to be there.

At an early presentation to the Lindner board, Sprague management sought funds for a much-needed expansion and upgrade of its silicon wafer processing facility in Worcester. Director Ron Walker began the discussion by commenting, "Well gentlemen, we have already spent a lot of money on your company, so I assume that in the future no more will be required." Upon hearing that the overall costs of expansion in Sprague Electric were such that it would be impossible to finance them without additional support from the parent corporation, Walker grumbled, "What a terrible business!"[3]

III: PROGRESS, THEN STRUGGLE TO SURVIVE, AND LIFE AFTER DEATH

The Worcester needs were eventually filled by two expansions of the original facility, and the 1984 acquisition of Solid State Scientific in Willow Grove, Pennsylvania. The Solid State acquisition proved a poor solution that would haunt the company for years. Even after detailed due-diligence by Sprague technical and marketing personnel had concluded that Solid State would do little either to solve the capacity requirements or to add appreciable new revenue, Penn Central management insisted on completing the acquisition. So Sprague Electric reluctantly agreed, sending Peter Loconto—at the time the Worcester operations manager—to try to make the most out of a bad situation. Performance improved but the losses continued for years.

At a multi-day management meeting in one of Penn Central's Florida properties, John Sprague had the unenviable assignment of telling the Sprague Electric story as the last person on the Friday afternoon agenda. For two days, executives had been sitting in a darkened conference room listening to hour after hour of presentations. Outside they could hear the surf breaking and smell the salt spray from the ocean, although there was little opportunity to enjoy any of it. When it was finally John Sprague's turn to make electronic components sound exciting, he could hear snores in the back of the room. No one seemed interested. Most wanted to get away, either to the airport and home, or to one of the many nearby golf courses, or to the beach. There were no questions, but as John

Sprague was walking out with the rest of the crowd, Carl Lindner commented, "Good job."

During a later corporate Christmas party, John Sprague and Carl Lindner sat next to each other. For all his wealth and success, Carl H. Lindner seemed a mild person, tall and distinguished with white hair, glasses, and a friendly smile. He often wore white suits, and enjoyed driving around his Florida holdings in one of his yellow Stutz Bearcats. It soon became apparent that Carl strongly favored businesses that had predictable sales and earnings, were non-cyclical, non-capital intensive, and weren't technology-based, the antithesis of Sprague Electric. Later he asked, "John, do you know what I am?" Sensing John Sprague did not know how to reply, he leaned forward and smiled, "John, I am the lender of last resort." Sprague wasn't exactly sure what that meant, but assumed it translated into something like, "buy low, sell high!" Unfortunately Sprague Electric didn't seem to fit that formula very well either.

The forecasted improvement in the economy did not occur in 1982 or even early 1983, as the severe 1981 economic recession dragged on. Then, with inventories at a dangerously low level, 1984 component orders surged and Sprague Electric scrambled just to catch up with the delayed demand. Unfortunately the surge was short-lived as the economy slowed again in 1985. It couldn't have happened at a worse time. Sprague Electric was in the middle of an aggressive and expensive expansion program just as sales dropped

III: PROGRESS, THEN STRUGGLE TO SURVIVE, AND LIFE AFTER DEATH

off a cliff. As good as the latter half of the 1970s had been, the middle 1980s were a disaster. The cyclicality of the electronic components industry requires a business model where fixed costs are maintained at a level that allows financial survival during the inevitable down cycles. However, following the extraordinary last half of the 1970s, and urged on by Penn Central, Sprague Electric continued to add fixed costs as if the good days would never end.

As Sprague Electric continued to decentralize corporate functions, move manufacturing to lower cost locations, and consolidate plants where it made sense to do so, the future of North Adams as the corporate headquarters had become increasingly uncertain. Employment there had steadily declined from a 1966 peak of more than 4000. Following the 1970 labor stoppage, it had dropped to just over 2000, and by 1982 was down to 1500. A Penn Central corporate visit to North Adams in the spring of 1983 sealed the city's fate.

The North Adams Harriman-and-West Airport has a single paved runway, lights but no tower. Sitting at the base of Mount Williams, it is subject to swirling wind turbulence. Its 4300-foot runway (it may have been shorter in the early 1980s) is adequate for handling jet aircraft, but even under good weather conditions, landing there can be challenging. One can imagine the anxiety of the passengers—a full load of Penn Central directors and executives—when they came into the valley on the Penn Central jet to land on what must have looked like a postage stamp.

This was a first time visit to North Adams for many of them. As the executives drove from the airport to nearby North Adams, they noticed the numerous old, often empty, textile mills. Upon entering the gates of the 87 Marshall Street complex, they were escorted up to the beautiful main conference room with its massive oak table covered with tan felt where Sprague Electric senior executives and board of directors welcomed them. Following cordial introductions, a limited number of presentations, and a brief tour (it appears that luncheon was skipped), the visitors rushed back to the airport, happy to be gone and hoping that the Penn Central jet had enough power to take off and clear the surrounding hills.

The feedback from Penn Central CEO Martinelli was swift and unequivocal. He made it clear that he was never going to fly into that airport again, that North Adams was no place for the headquarters of Penn Central's highest technology investment, and that he wanted headquarters moved to somewhere appropriate, such as the Route 128 area of Boston. As a result planning began for a new world headquarters in Lexington, Massachusetts, the restructuring of the Sprague board of directors and management, and an expensive initiative called Actions for Profitable Growth.

Actions for Profitable Growth was announced in early 1984 after nearly a year of detailed planning by the Penn Central Strategic Management and Planning Group, Gene Swartz, and Sprague management. It set seemingly impossible goals of reaching net sales of $1 billion and pretax earnings of $100 million by 1987. This meant rapid expansion of all the existing Sprague Electric product lines and probably strategic acquisitions as well. The only possible way such a plan could succeed was if the electronic component market enjoyed another steady growth period similar to the latter half of the 1970s. This did not happen.

Although full occupancy and the formal dedication of Sprague world headquarters at 92 Hayden Avenue in Lexington did not occur until 4 December 1984, personnel transfers and hiring began midyear. In announcing plans for relocating headquarters to Lexington in early 1984, CEO John Sprague said that related transfers from North Adams would total only about a dozen people. However, when all was said and done, consolidation of most corporate functions in Lexington and in a new distribution facility in Mansfield, Massachusetts, coupled with acceleration of the corporate-wide decentralization process, led to an actual loss of more than five hundred and fifty hourly and salaried jobs in North Adams over the next several years. The company badly mishandled explaining the difference, so that when the new number was finally released in October 1984, the news stunned and infuriated the North Adams community, Sprague Electric's North Adams employees, and new Mayor John Barrett, creating a feeling of betrayal. Although more than 2500 Sprague jobs had already left between 1970 and 1984 (the majority of those resulting from the strike) and although some manufacturing would remain in North Adams, Sprague Electric was perceived as having abandoned the city that had helped make it great. In actuality, by mid-1987 there were still some six hundred and forty Sprague jobs in North Adams, mostly in Filters, Wet and Foil Tantalums, and at Commonwealth Sprague in Brown Street, making Sprague Electric the second-largest Berkshire employer (excluding GE-Pittsfield), behind Williams College (with eight hundred and thirty).[24] Unfortunately by 2000, eight years after Sprague Electric had ceased to exist as an entity, most of those jobs were gone as well.

The new headquarters was located at the intersection of Route 2 and Route 128 west of Boston in a two-story former Burroughs office building that had been completely gutted and rebuilt. The building was of grey brick and glass and included a wide second-story recessed balcony overlooking Route 2. Sprague Electric had finally gone big time, at least as far as Penn Central was concerned.

Although the move from North Adams seemed appropriate for a soon-to-be billion-dollar Company, it made little sense considering the upheaval of key personnel, and especially when viewed strictly from a cost standpoint. Consolidation in the Research Center, or in a new facility in North Adams' Hardman Industrial Park, had been briefly considered by the planning group, but quickly abandoned because it would not have solved the labor problems, and because Penn Central management clearly wanted a metropolitan location. Ironically, less than three years later, the new headquarters facility in Lexington would also be closed, and there would be a new company, Sprague Technologies, Inc., with a new headquarters in Stamford, Connecticut, and a new CEO.

Nonetheless, everything was upbeat at the dedication of the new headquarters as John Sprague addressed the invited guests. "With the support of PCC we have formulated a far reaching accelerated growth plan having a goal of $1 billion in sales by 1987. Current 1984 estimates indicate we will reach record sales in excess of $500 million (the actual was $571 million) and we are confident that we can reach our ambitious goal for 1987." Al Martinelli followed by describing Sprague Electric as the "flagship of the Penn Central Company."

John Sprague also introduced his new, decentralized organization built around a group of world-wide product managers, each with full worldwide profit and loss responsibility for their product groups. Reporting directly to John Sprague in Lexington were Sprague Electric veterans Executive Vice President Don McGuiness (operations), Senior Vice President John Murphy (technology and support), and Fred Windover (Vice President and Chief Counsel), along with newcomers Senior Vice President Jack Darcy (marketing and sales), Senior Vice President Larry Switzer (finance and administration), and Doug Smith (planning). The world-wide product managers were all seasoned Sprague Electric executives: Tom Browne (Aluminums and Thick Film Networks), Peter Loconto (Integrated

Circuits), Peter Maden (Tantalums), Hal Mahar (Special Components), Bill Milton (Ceramics), and Jim Sherry (Films). Yet what seemed so logical and clean on paper unfortunately turned out to be far less effective in reality.

The problems began in Lexington where John Sprague was never able to create a cohesive team, primarily because there were too many personal agendas. Although Jack Darcy's aggressive and profane persona rubbed some of the older Sprague sales veterans the wrong way, he and John Sprague hit it off immediately, and Darcy brought a wealth of experience, especially in the electronic distribution industry. Larry Switzer was a different matter. A former Iowa State football standout, Switzer was a big man, both figuratively and physically. It quickly became apparent that what he really wanted was the CEO job and the sooner the better. After John Sprague learned that Switzer had taken his case directly to Penn Central management, Switzer was gone quickly. Following several interim appointments, Switzer was replaced by Sprague veteran, Don Christiansen.

All the world-wide product managers reported directly to Don McGuiness, making him the clear number two person in the company. He had done an outstanding job in semiconductors, but shortly after his promotion he told John Sprague that it was time for Sprague to step aside, become a figurehead, and let McGuiness run the show. John Sprague had no intention of retiring, and relations between them quickly soured. McGuiness was a firm disciple of Erhard Seminars Training, which claimed "to transform one's ability to experience living" and he insisted that his new subordinates also embrace the discipline. Most objected, and after a year Don was also gone and not replaced, as John Sprague chose to assume his operating responsibilities personally.

Peter Maden was particularly effective, in no small part because he was responsible for Solid Tantalums, one of the few capacitor families where Sprague Electric had been able to maintain worldwide market leadership since its invention by Sprague in the early 1950s. At the other end of the spectrum, Tom Browne's change of allegiance when Ed Kosnik first appeared on the scene was so swift and so complete it almost seemed as if he had never worked for John Sprague at all.

Even if Lexington had operated perfectly, it probably would have only marginally changed the outcome. For nearly ten years, through up-and-down market cycles, expansions and contractions, changes in management and organization, Sprague Electric had operated successfully. The Semiconductor Group had become an important asset rather than a liability. After five successful years during the latter part of the 1970s, even during the downturn which followed, from 1981 through 1984, Sprague had maintained profitability. When "Actions for Profitable Growth" became the new business model, however, it failed miserably.

As expenses soared, John Sprague knew that the company was heading the wrong direction. Although the move out of North Adams made long term sense, at the time Sprague Electric could not afford it. The same was true with Actions for Profitable Growth. Instead of trying to grow everything, Sprague Electric should have concentrated on those businesses at which it excelled. Despite this, John Sprague decided to proceed with the plan. In looking back years later he said he felt as if he were on a train platform, and the express was about to leave, and if he didn't climb aboard, he would be left behind. There was also an important evolving competitive problem. Starting with consumer electronics—a key market for many Sprague Electric product families, including integrated circuits and aluminum capacitors—Japanese corporations were becoming formidable world-wide competitors. They were beginning to control the multi-layer ceramic capacitor industry as well. As off-shore competition intensified, tantalum capacitors were increasingly the only dielectric family where Sprague Electric continued to maintain a leadership position.

The year-to-year volatility of revenues was driven primarily by economic cycles. The brief 1984 surge in components was followed in 1985 by a plunge in incoming orders and revenues. Although IBM was Sprague Electric's largest and most important customer for much of its life, there were problems doing business with the world's largest computer manufacturer. IBM found accurate forecasting of its component requirements and inventory control difficult. In up markets, Sprague was often pushed to the limit to meet IBM's orders, only to face sudden unexpected cancellations

III: PROGRESS, THEN STRUGGLE TO SURVIVE, AND LIFE AFTER DEATH

when the economy dropped. The solid tantalum and networks business units were particularly savaged by this problem.

Penn Central Management circa 1985. Left to right: Carl Lindner, Al MArtinelli (sitting), Ed Kosnik, Pug Winokur. (Penn Central Annual Report)

The financial losses in 1985 and 1986 resulted on the operating side from lower revenues and from added overhead costs related to Lexington and the actions for profitable growth initiative. However, those operating losses were overshadowed by restructuring costs and reserves. For further detail of the Sprague Electric operating performance during this period see Table 7 in Appendix 1. Growth at any cost quickly shifted to survival.

The operating losses were concentrated in Solid State Scientific, Aluminums (in 1985), and multi-layer ceramic capacitors. In Sprague's core fixed-capacitor segment, only tantalum capacitors (both solids and wet & foils), showed continuing major profit potential, along with screened thick film networks, which were used primarily in computer applications. However, the volatility in the order patterns of IBM was a major risk there as well. Even during 1985 and 1986, Worcester's integrated circuits and Concord's discrete semiconductors plants both remained modestly profitable.

By 1986, Carl Lindner had assumed complete control of Penn Central. He had eliminated most of the prior Penn Central management, and moved the Penn Central headquarters to his home city of Cincinnati. As he continued to redeploy and shed parts, having already created Sprague Technologies, Inc. with Ed Kosnik as CEO, in August 1987 he spun it off to the Penn Central stockholders as a separate company. However, Sprague Technologies was scarcely independent. The board of directors was a group of Penn Central loyalists, Lindner was its chairman, and he controlled more than one third of its stock. It soon became clear that he wanted to exit all of Penn Central's manufacturing businesses, including Sprague Electric, and use the generated cash to expand his property and casualty insurance interests.

Penn Central brought the consulting firm McKinsey to Lexington to "help," and after an expensive year on site, the consulting firm concluded that the sum of Sprague Electric's business units as individual entities was worth more than the company as a whole. In other words McKinsey recommended breaking up the company and selling off the pieces.

While the agony of downsizing was consuming the company in 1985 and 1986, John Sprague attempted to visit every worldwide location that was being closed to try and explain what was happening. It was a dreadful task because at stake were not just bricks and mortar, but people's lives. Between 1984 and 1987 Sprague Electric's worldwide employment dropped from 12,000 to 8,400. On at least one occasion John Sprague was not sure he would get out alive after addressing the angry employees of an about-to-be-closed plant in West Germany.

As Sprague Electric was imploding in 1985 and 1986, Penn Central management decided that they needed to find a way to ease John Sprague out of the picture, especially if breaking up the company and selling off the pieces was one of the major options being considered. Initially they created a two-layer corporate structure in which John Sprague continued as president of Sprague Electric, but now reported to former Penn Central Executive Vice President and Chief Financial Officer Ed Kosnik, CEO of Sprague Technologies, Inc. Initially John Sprague was certain that Kosnik's primary role was to pretty-up the company for sale either whole or piecemeal. However, over time John Sprague came to believe that Kosnik hoped to create and lead a new and successful Sprague Electric. He might have succeeded with a company built around the tantalum dielectric system (solids and wets and foils), networks, and—depending on his willingness to take on the challenge—the Sprague Semiconductor Group. However, he never had a chance.

At a well-attended spring 1987 sales meeting on Cape Cod, which also included spouses, Kosnik celebrated "A New Beginning" for the sixty-year old Sprague Electric Company. Although John Sprague was still titular president of the company, his sole responsibility was to present an abbreviated history of Sprague Electric. In the middle of his talk, John became suddenly and uncontrollably overcome by what was happening to the company founded by his father and where he himself had worked for twenty-eight years. He saw the shocked audience reaction and heard his wife's strangled "Oh John," and noticed that Kosnik, seemingly unmoved, was staring stonily straight ahead. Chagrined and deeply embarrassed,

III: PROGRESS, THEN STRUGGLE TO SURVIVE, AND LIFE AFTER DEATH

John recovered, and after apologizing, was able to finish without further incident. On the way out several well-wishers stopped to say he had no reason to apologize. Nevertheless he had given Penn Central management the opening it needed.

Weeks later John Sprague was given his walking papers by the Sprague Board: "John, it is clear that you just aren't with the new program." He was in no position to argue, and felt the severance package offered was more than fair. It included a position as vice chairman of the new Sprague Technologies Board (Carl Lindner was Chairman), an office and consulting assignment in Worcester for a year (a difficult arrangement, to say the least, for semiconductor world-wide product manager Dick Morrison), and help in creating a consulting company, John L. Sprague, Associates, which was incorporated 15 April 1988 "to provide consulting services to technology based firms."

In early 1987, the Sprague family suffered two overwhelming personal tragedies. On 10 April 1987, Bob Sprague, Jr. died in a fiery crash at the North Adams Harriman-and-West Airport. Unable to cope with the loss of her oldest son, Florence Sprague died of heart failure on 1 June. Robert Sprague grieved, but his spirit was not broken by these twin tragedies. Rather than lament these losses and the dismantling of the company he had founded, Robert Sprague chose to enjoy his remaining family, which at the time included eight grandchildren and eight great grandchildren, along with his other interests such as the Elm Tree Foundation and especially the Williamstown Theatre Festival, both of which he helped found. Even long after he had withdrawn from the electronics Industry, he continued to receive honors. In February, 1985 he was elected to the National Academy of Engineering, one of the highest professional distinctions an engineer can receive in the United States. He lived another four years and died peacefully in his sleep at home in Williamstown on 28 September 1991.

Sprague Technologies, Inc.

In 1987, Sprague Technologies returned to modest profitability with net income of $3.6 million on sales of $437 million. Although Penn Central was no longer available to serve as STI's bank, Kosnik reassured stockholders in the 1987 Annual Report that "internally generated funds (plus a recently negotiated $40 million revolving line of credit) should be sufficient to support STI's needs in the short term." However, following another market collapse that began in late 1987, trouble was again on the way. For the whole of 1988, STI had net income of $13.2 million on sales of $505 million. But the improvement over 1987 occurred entirely in the first six months as the recession finally caught up with the electronics industry during the middle of 1988. It proved a deep recession that lasted into the early 1990s. Internally, the search began again for a

III: PROGRESS, THEN STRUGGLE TO SURVIVE, AND LIFE AFTER DEATH

new exit strategy. Merger discussions started with AVX, Kemet, and several other potential partners, but did not come to agreement. Preparing for a probable separation into two companies, Sprague reported results by product segment for the first time in 1988: Components (basically all the non-semiconductor products) and Semiconductors. The former had sales of $386.5 million and operating income of $29.0 million, while Semiconductors (including Solid State Scientific) broke even on sales of $118.1 million. For STI as a whole, profits continued to be dominated by solid tantalum capacitors. Nonetheless, STI was heading into a classic downturn similar to the one that had just ended.

As the recession continued into 1989, STI was back to restructuring and moving towards being solely a passive component company. $344 million in revenue led to a loss of just under $100 million, of which $80 million was another restructuring reserve. All business units, including Solid Tantalums, were savaged by a $40 million IBM shortfall, once more caused by massive excess inventories at the customer level. Only tantalums were able to eke out a modest profit. Multi-layer ceramic capacitors, Fil-Mag (the Filter Division, which manufactured electromagnetic filters and magnetic components), semiconductors, and several other small business units were discontinued, while capital expenditures were limited almost entirely to surface-mount technology (the growing trend in device packaging), to networks, and solid tantalums. Considering the 1989 restructuring reserve of nearly $100 million, at first glance 1990 sales of $315 million and a loss of $49 million looks worse than it was. Nonetheless, things seemed to be looking up now that STI was a supplier of only tantalum and aluminum capacitors and thick film networks. On 19 December 1990, the semiconductor business was sold to Sanken Electric Ltd. of Japan (and renamed Allegro MicroSystems) for $61.9 million in cash, versus a book value of $106 million. Therefore the STI loss for the year was almost all from the book loss on this transaction. With all bank offerings repaid, cash and equivalents of nearly $33 million, and a $160 million tax loss carry-forward, Kosnik reported that STI was now well positioned for future growth and profitability, despite the continuing recessionary environment.

By 1991, Kosnik's annual report message had changed dramatically:

> "In recent years Sprague divested unprofitable business units in attempting to create a profitable core component business. Despite downsizing, the Company was unable to produce consistent profitability as the industry matured and the component recession lingered. The divestitures culminated with the sale of the Tantalum Capacitor and US Thick Film Network businesses in February 1992 (to Vishay Intertechnology, Inc. for $120 million in cash plus 'certain other considerations'). With the sale the company has effectively exited electronic component manufacturing."

On 17 April 1992, the STI Board of Directors elected Chairman Carl Lindner to the additional post of CEO, replacing Ed Kosnik. All that remained of Sprague Electric was approximately $100 million in cash and a tax loss carry-forward in the range of $130 million. Adding this to what had already been accumulated within Penn Central by prior sale of other businesses, Lindner continued the Penn Central transformation into an insurance company, which in 1994 was renamed American Premier Underwriters, Inc.

Although Sprague Electric no longer exists, the successor businesses of its units and operations are spread around the world. These include the previously mentioned MRA Labs, two specialty private film capacitor companies (SBE, Inc. in Barre, Vermont, and Dearborn Electronics in Longwood, Florida), and a private aluminum capacitor company, Barker Microfarads in Hillsville, Virginia. United Chemi-Con also continues to operate the former Sprague aluminum capacitor plant in Lansing, North Carolina. All proudly acknowledge their Sprague Electric heritage.

Vishay Intertechnology, Inc purchased the Sprague tantalum capacitor businesses in 1992. Founder and CEO Dr. Felix Zandeman, who died in June 2011, moved much of the solid tantalum production to Israel, drawn there because of lower labor rates—at least compared to the United States and Europe—and strong work ethic. Zandeman was also successful in negotiating tax incentives with the Israeli government. He had also acquired the Sprague trade name, and today one can purchase a broad line of Vishay-Sprague capacitors, dominated by the solid tantalum surface mount configuration that Sprague had pioneered prior to the acquisition. The industry move toward automatic attachment to complex circuit boards of all types of active and passive components was driven by the need for more complex and dense electronic circuitry.

Succeeding where Sprague Electric had failed, Vishay Intertechnology, with revenues in excess of $2.0 billion and more than 22,000 employees worldwide, became one of the world's largest suppliers of passive components, while also having a limited offering of semiconductor devices.

Vishay consolidated wet and foil tantalums plus some solids in Sprague's former Sanford, Maine plant, leading to major job losses in Concord, New Hampshire and Tours, France. Using market share as a measure, these moves have apparently not been completely successful. Prior to the acquisition Sprague shared the world tantalum capacitor market lead with Kemet at roughly twenty-five percent each. According to a Paumanok Publications, Inc. March 2008 report, Vishay's Tantalum share had slipped to ten percent, behind AVX, Kemet, and NEC/TOKIN.[25]

The other Sprague business that succeeded admirably was Allegro MicroSystems, LLC, the former Worcester-based Sprague Semiconductor Group purchased by Sanken Electric Co., Ltd. of Japan in 1990. At the time, revenues were stagnant and profitability was minimal. Unlike Penn Central or probably any other potential U.S. acquirer, Sanken proved to be a patient owner. Each successive CEO: Dick Morrison, Allan Kimball, and today's President and CEO, Dennis Fitzgerald, contributed to the company's current successful financial position. Annual revenue as of 2014 was approximately $500 million, and operating margins exceeded fifteen percent. Allegro's operating performance continues to be ranked in the upper second quartile of analog semiconductor companies. With worldwide employment exceeding 3,000, Allegro is headquartered in Worcester, Massachusetts (U.S.A.) with design, applications, and sales support centers located worldwide. Allegro built itself around two basic product families: Hall-effect magnetic sensors, and power integrated circuits, both pioneered at Sprague Electric in the 1970s. Allegro has succeeded by sticking to and capitalizing on its greatest strengths. Automotive components are a particularly important market, comprising more than two-thirds of the business. A luxury automobile might contain more than ninety magnetic sensor, motor driver, and regulator integrated circuits. The average Allegro content per worldwide vehicle is greater than $4.00.

If so many former Sprague Electric business units exist today as successful parts of other corporations or private companies, why did the original parent company fail and disappear as a single entity? It would be easy to blame Penn Central. However, the seeds of Sprague's demise were laid years before when Sprague Electric was first acquired by General Cable and ceased to be in charge of its own destiny. Three primary causes led to this acquisition. First, the long, tortuous, and expensive process of becoming a successful niche supplier of semiconductor devices sapped the company of too much of its financial, managerial, and technical resources at a time when it could least afford it, as foreign competition, especially from Japan, intensified. While Allegro proved that Sprague could have succeeded in semiconductors, Allegro's success came only after a series of expensive false starts and strategic errors.

Secondly, because of the concentration of resources in semiconductors, Sprague Electric was never able to create a viable position in multi-layer ceramic capacitors, even as a niche supplier. Finally, the increasingly acrimonious labor relations in North Adams that led to the devastating ten-week 1970 strike nearly destroyed the company, causing major losses both financially and in capacitor market share, and making a weakened Sprague Electric vulnerable to acquisition only a few years later.

Ceramics, particularly multi-layer ceramic capacitors, dominate the present capacitor market because the related technology continues to create components that can replace applications previously served by other dielectric families at a fraction of the cost. It is a highly competitive market and, except for high-value specialty niches, the average price for the entire mix of ceramic capacitor products is less than a penny a unit ($0.0069 in 2007). Yet there remain a number of successful "U.S." capacitor suppliers, including Vishay, with a broad mix of both passive and active devices, U.S.-based Kyocera/AVX in ceramics and tantalums, Kemet in tantalums, and a host of smaller specialty firms. Given a chance, Sprague Technologies might have succeeded as a tantalum capacitor and networks supplier if Penn Central hadn't pulled the plug. Still this is wishful thinking because Carl Lindner never wanted to be in that type of business.

Life After Sprague Electric and the Deindustrialization of North Adams

When Sprague Electric completed its departure in the 1980s, and the city of North Adams desperately searched for a replacement employer, there was none in sight. After nearly one hundred years of being cared for primarily by a single dominant employer, the absence required an agonizing adjustment for the city and its residents. Even with Arnold Print, then Sprague Electric, and now the Museum of Contemporary Art, the North Adams population declined steadily from a peak of 24,200 in 1900 to a low of 13,689 in 2011. Help in trying to reverse this trend came from an unexpected source.

The Massachusetts Museum of Contemporary Art (MASS MoCA)

87 Marshall Street, former home of Arnold Print Works, Sprague Electric, and MASS MoCA (courtesy of Paul Marino)

A 1987 visit by Thomas Krens to the Schaffhausen Museum near Zurich, Switzerland, once a textile factory and now a contemporary art museum, was the early inspiration for Massachusetts Museum of Contemporary Art. Krens was the charismatic director of the Williams College Museum of Art, and had been considering several of the nearby abandoned North Adams mills as additional space for the Williams College Museum's growing collection. Schaffhausen suggested an additional intriguing possibility, that the North Adams mills be used to house massive contemporary art exhibits for which, at the time, little appropriate space existed in the United States.

III: PROGRESS, THEN STRUGGLE TO SURVIVE, AND LIFE AFTER DEATH

Encouraged by North Adams officials, especially Mayor John Barrett, the former Sprague Electric facility at 87 Marshall Street quickly became the leading candidate. Covering thirteen acres, the sprawling twenty-six building complex offered nearly one million square feet of potential museum and economic development space – an almost unimaginable opportunity. On the other hand the facility was deteriorating rapidly, and with dozens of interconnecting bridges and tunnels, transforming the property into a museum seemed nearly impossible.

Krens' initial concept was vast, with an estimated cost of $72 million. In addition to museum galleries, it would include a conference center, restaurant, small hotel, additional commercial and industrial space, and a river walk (a la San Antonio). An estimated six hundred new jobs would be created, some $21 million additional revenue brought into the economy, and local and state tax revenues increased by more than $1 million a year. Massachusetts Governor Michael Dukakis also was a strong early proponent, because North Adams had yet to participate in the surging "Massachusetts Miracle" of the late 1980s.

In 1988 Krens left for the Guggenheim Museum in New York City and his assistant, Joseph C. Thompson assumed the role as director of the project, eventually becoming director of the museum, a post he still holds today. In December 1993 funding responsibility shifted to a new entity, the MASS MoCA Foundation, and MoCA's mission was gradually repositioned from "the world's largest contemporary art museum" to one where the museum would become an international, multi-disciplinary cultural center incorporating dance, music, theatre, history, education, and technology in addition to contemporary art.

Despite strong local support, including key early financing by Williams College, it was touch and go, especially in obtaining the necessary financing from the Commonwealth of Massachusetts. Still Thompson and his team prevailed and on May 29, 1999 MASS MoCA officially celebrated its official opening with a lavish black tie gala attended by more than 1200. Open to the public the following day, many visitors were former Sprague Electric employees who couldn't believe what had happened to the facility where they formerly manufactured electronic components. More than one was heard to comment, "I wonder what 'R. C.' would think." He would have been pleased!

Silicon Village

While MoCA was being created, additional exciting opportunities appeared and by 1999, some were calling the Northern Berkshires "Silicon Village."[26] Williams College served as the incubator when in 1992 two classmates, Bo Peabody and Brett Hershey, joined with their economics professor, Richard Sabot, to form Tripod, an early on-line social network offering advice to students for a fee. Before long it also offered a more important service as a free portal for creation of web pages.

III: PROGRESS, THEN STRUGGLE TO SURVIVE, AND LIFE AFTER DEATH

Successful beyond anyone's original expectations, in 1998 it was bought by Lycos for approximately $58 million and—except for the three founders—its forty or so Williamstown-based employees were soon moved from North Adams to the Lycos facility in Waltham, Massachusetts, or terminated.

In January 1997, Berkshire Capital Investors (BCI) was formed with Williams College alumni George Kennedy, Taylor Briggs (now deceased), and Robert McGill as founding partners. Buoyed by $3 million in early seed money from Williams College, creation of a number of start-ups followed, with such as exotic names as Streetmail (later Everyday Health), Resounding Technology, Geekcorps, Eziba, PRG (later Boxcar Media), Emergon, VoodooVox, Xtend Energy (later CPower), Healthshare, and Skillview. Some located in MoCA while others were in downtown North Adams and Williamstown. BCI was the financial catalyst for many, as was its affiliate, Village Ventures, co-founded in 1999 by Bo Peabody and Matt Harris. As the recession of the late 2000s deepened, the pace of new company start-ups in the Northern Berkshires slowed dramatically, and in recent years neither BCI nor Village Founders has made new capital investments in the area. In early 2012, the *Berkshire* Daily reported that Village Ventures was closing down its Williamstown operation. Although a few technology companies remain and are thriving in the Northern Berkshires, unfortunately "Silicon Village" was mostly a public relations creation.

The Future of North Adams

With Sprague Electric now gone and no equivalent replacement in sight, the local economy continues to deteriorate, despite the increasing success of MoCA and improving stature of the Massachusetts College of Liberal Arts (MCLA). Different studies and forecasts portray a once dynamic industrial economy now dependent on arts and tourism and a region continuing to lose population as it moves towards becoming a collection of retirement communities. There are successful businesses such the venerable Crane & Co., a world leader in premium cotton cloth paper that employs 300 in North Adams, and smaller businesses in publishing (Storey), information technology (Boxcar Media), back office customer service (Streetmail), and energy (iPowerPlay). Unfortunately their collective new job creation so far has been insufficient to replace Sprague Electric's departure.

The only entity in the Northern Berkshires with sufficient resources to reverse this trend is Williams College, and this would only be true if Williams were to expand its current role as one of the world's finest small liberal arts colleges by creating an engineering school in the natural sciences. Apparently there is no way this will ever happen unless the local economy begins to create a problem for the college's ability to hire the finest educators, coaches, and support personnel.

EPILOGUE: JOHN SPRAGUE'S REFLECTIONS ON ROBERT SPRAGUE

Dad

When I was a little boy growing up in Williamstown in the 1930s, my father travelled almost continually for the struggling new business he had founded in 1926. Upon arriving home at night, exhausted from a week on the road, he still always had some present for me in the top of his battered suitcase. I don't remember what they were, but I still recall my excitement as, with a twinkle in his eye, he slowly opened the suitcase, and the joy of spending even a few minutes with him as he talked about his trip. To me he was the most wonderful and smartest man in the world.

In the early years his influence was more subtle than verbal. Although we seldom talked about life, or morality, or what I wanted to do, we did things together, particularly sports. About five feet, ten inches, and very fit, he was a good athlete (at Annapolis he was a gymnast on the flying rings), but also a perfectionist. He tried golf, but dropped it when he couldn't break 100. We played tennis for years, but when as an early teenager I beat him in singles for the first time, from then on we only played doubles and then as ferocious and generally successful partners. Skiing together started when I was seven and he was thirty seven (he was still skiing in his 80s). When he could no longer keep up, he decided to write a book on the physics of skiing, *Parallel Skiing for Weekend Skiers* (my wife, Jid, was his only pupil, with mixed results). There is a novice run at Carinthia, now part of Vermont's Mount Snow, where he used to practice making perfect turns on a shallow slope groomed absolutely flat. Steep pitches and moguls provided a much greater challenge, so he just didn't ski them anymore.

My father liked people, so winter ski trips in the 1940s and 1950s to Mont Tremblant in Canada (my mother never came) found our initial dinner table of four or so at least tripled by the end of the week, with my father clearly on center stage. After dinner, if there was a small orchestra, he was soon on the floor, dancing with the prettiest woman in the group.

When I was thirteen, my parents sent me away to Middlesex School in Concord, Massachusetts. I was miserably homesick the first three years, but no matter what ploy I tried they forced me to stay. Having displayed an early interest in chemistry, as college decision time approached I asked my father about Annapolis, where both he and my grandfather had received excellent educations. Here he was very firm, "If you want to study chemistry the United States Naval Academy is not the right place for you." So I decided on Princeton because that is where my favorite Uncle Hugh (my mother's brother) had gone. It was a good choice.

During my academic career I maintained a good, but never brilliant, B to B+ average. My best work by far, both in college and later graduate school, was in the laboratory. My parents never bugged me about my grades, even when they started to deteriorate as I began to spend increasing time travelling back and forth to Smith College (it wasn't all for naught since most visits were to see Jid Whitney and we now have been married more than sixty-two years). My father never seemed too worried; with his lifetime optimism, he knew "John will get it done!" I wasn't so sure, but did, just squeaking by at Princeton as an outstanding research thesis saved the day.

During my first year of sea duty on the *U.S.S. Kleinsmith*, I had the opportunity to see my father in Washington, DC as he was being vetted for Under Secretary of the Air Force under Eisenhower. I was completely surprised how at ease he was in this environment (and that my mother wasn't) and to realize that he felt destined to play on a much bigger stage than Sprague Electric. Soon afterwards I applied, and was accepted for, flight training at Pensacola. When I came home to Williamstown on leave before reporting to Florida, my family descended on me to try and change my mind. Jid and my mother were convinced I would be killed during carrier training, my brother (who taught me to fly) said I was crazy, but it was my father who quietly asked the key question, "Do you want to spend the rest of your life flying airplanes for a living?" "No, but right now I really want to fly jets off of carriers." Not to be. So I reluctantly cancelled my orders and, with my father's urging (and help?), applied for and was accepted at the United States Navy's Officers Electronic Materials School in San Francisco. There my wife and I fell in love with California and I began my education as a semiconductor scientist, a life-changing decision with later profound impact on me, my family, and on Sprague Electric.

When my three years of active duty were completed Jid and I spent the summer of 1955 in Williamstown before heading to California where I had been accepted at Stanford University as a PhD candidate in chemistry. I had already decided I didn't want to work at Sprague Electric (I have always had a philosophical aversion to nepotism), and I knew that a doctorate would greatly increase my hireability. My father had other ideas; planning years ahead, he apparently hoped I would one day head the company. In the fall of 1955 my father and I drove across the United States. Jid, who was pregnant with our first child—John— flew out later. The trip was just the two of us spending almost a week together. It was a delight in many ways, but for the first time I experienced the full force of his personality as a salesman. "After graduation I hope you will consider coming to work at Sprague electric." Later, as semiconductors became my passion, he added, "I need you".

Because my thesis was on semiconductor materials, I was given complete access to the Sprague Research Laboratories, advice from some of the best minds in the industry, electronic measuring equipment, and even evaluation of the crude devices I was making in my basement lab at Stanford.

In early 1959, as I approached graduation, the die had already been pretty well cast. To my chagrin, despite an important thesis in the new field of semiconductors, none of the west coast start-ups would hire the son of electronic component giant Robert C. Sprague. There were attractive job offers in other fields, but if I wanted to work in semiconductors, Sprague Electric was the only choice. So reluctantly putting philosophy aside, but also excited to be working for one of the industry's true geniuses, Dr. Kurt Lehovec, I returned to North Adams in the spring of 1959 as a research scientist, exactly as my father had wanted.

R. C.

Being the CEO of a public, high growth, technology company is one of the world's most exciting, risky, and demanding responsibilities, and successful ones are worth their weight in gold (which often is what it takes to hire them). The pressure is unrelenting; he or she must simultaneously balance the needs of multiple constituents, including customers, employees, communities, suppliers, regulators and stockholders, all while continually trying to stay ahead of the competition. Some do it for the power, the perks, and the compensation, while others believe they can really make a difference. My father was one of the latter.

When I started my research in the Marshall Street complex, although we were still many organizational layers removed, my father was now also my boss. So I had to view him from a completely different perspective. Almost everyone I met revered, respected, and trusted him as the person who had done, and would continue to do, the right thing for the company and therefore for them. As with almost any boss, some also feared him because of the power he had over their lives. Sprague Electric was like its own community within the city of North Adams as a whole, with sports teams and music groups and banquets and other gatherings, which he attended whenever possible. Although, as a public company, financial results were readily available, Robert Sprague still used the *Log* to interpret them to the employees and regularly used this internal publication to communicate how the company was doing. While often criticized by union labor representatives who accused him of using this "fatherly" approach only to keep wages low, he really cared about his employees and felt it just made good business sense for them to feel they were in good hands with "R. C." at the helm.

He surrounded himself with excellent people who he felt complemented his own strengths and weaknesses, and over the years used a series of executive committees, planning committees, technical advisors boards, and the like to create the form of consensus management that he used to run the company. However, consensus usually meant agreeing with what he had already decided.

He exhibited one characteristic at work that I had almost never seen at home, occasionally exhibiting flashes of real anger when he was tired or when he felt he was faced with pure stupidity. Warned by his secretary, these were times to stay away from him and let him stew in peace.

The 1947 invention of the transistor revolutionized the electronics industry, and clearly Sprague Electric needed to be in that business. Robert C. Sprague saw this as the next great step for Sprague Electric. Although by the mid-1950s the preferred industry technology was the batch-process MESA-junction transistor, Sprague chose Philco's highly mechanized electrochemical process which—using germanium—at the time produced the world's fastest switching transistors one-at-a-time. However they could never be integrated into a single substrate, and the planar process required silicon. I believe my father made this decision solely on his own (at the time there was still no one in the R & D organization to advise him otherwise). I guess the reason was that he didn't want to be part of the crowd, speed was king, and he completely underestimated how fast semiconductor technology would move compared to the capacitor industry. So Sprague Electric started down the wrong path with a tidy little business that disappeared within a few years.

In the meantime in the early 1960s, a small team in the North Adams R & D center developed an industry-competitive planar process and related family of devices (see Appendix 2) and with the hiring of Bob Pepper in 1964 created a pilot manufacturing facility and began design of a family of linear integrated circuits, and later, power integrated circuits that were to be transferred to Concord, New Hampshire and become the basis of a profitable and growing specialty semiconductor business. Now firmly on the planar band wagon, once more Robert Sprague was in a hurry and committed to building and staffing the Worcester, Massachusetts semiconductor plant.

The argument we had over changing to a niche semiconductor strategy was one of the low points of my life. I had to convince the man I loved and respected most in the world, who starting from scratch had built one of the world's most successful passive component companies, that this time he was dead wrong. He only made things much worse by calling me a "coward", inferring that, were he in my position, he would have the ability and will to make his broad-based second source strategy successful. I prevailed, from then on he backed me completely, but even as semiconductors became profitable the company never completely recovered from the damage done in getting there.

I am certain that, at least in his business life, the low point for him was the 1970 ten week strike. Convinced he had made Sprague Electric a good place to work and compensated his employees fairly, he couldn't imagine that they would take such an action, endangering not just the life of the company, but also their own. Yet they did and, never afraid to face a fight head on, as he daily crossed the increasingly raucous picket lines the bond he and other members of management had felt with the city of North Adams gradually began to crumble. And in a difficult economic environment and with customer commitments that had to be met, having no other choice he began the movement of 2000 jobs out of the city, a movement that only accelerated as the company moved toward decentralization and was acquired a few years later.

Ever the optimist, my father had the wonderful capacity always to look forward and very seldom back. When, in a vote of no confidence by the Board of Directors, he was replaced in late 1971 by Neal Welch as CEO and Chairman, he pressed on as Honorary Board Chairman, Executive Committee Chairman, and still largest stockholder. He attempted to grow the company using the Mostek Model of minority investments in high technology start-ups, one in audio speakers, another in electron-beam addressable memories, and a third in liquid crystal displays. Unlike Mostek, none succeeded. He expanded and diversified the Board, only to end-up at war with the independent Directors over staying in semiconductors and over who was to lead the company. A potential acquisition by Cabot Corporation fell through, but—having found a kindred spirit in Bob Jensen—at the end of 1976, Sprague Electric became a wholly owned subsidiary of General Cable with Neal Welch as Chairman and CEO and I was named President and COO. Although Robert Sprague's responsibilities were reduced to Honorary Chairman, he attended all the Board meetings, as was his habit, read every word of the voluminous Board Books, and asked only intelligent questions. The sole tension I ever saw between him and Jensen occurred when he mistakenly left his Board book in an elevator on the way to a meeting. It was never found, and Jensen was rightfully furious. Although I was now CEO and President, Robert Sprague's official association with the company ended when it became a subsidiary of Penn Central. After recovering from the depression of the strike, during this entire period of furious change, questions of his competence, and eventual loss of his company, I can never remember him being down, or complaining, or even questioning past decisions. Like most great leaders Robert Sprague had done what he believed was correct, readily embraced change however difficult that might be, and—if mistakes were made—it was never for lack of trying or being unwilling to give his all. Even as Sprague Electric and I struggled through the 1980s, he chose to never get involved, give advice, or criticize my decisions. He was always looking forward, at least until 1987 when my brother, Bob, was killed and my mother died.

Elder Statesman

During his business life my father was actively involved in two trade associations. The first was the Associated Industries of Massachusetts of which he became a director in 1945 and President in 1951. His related comment in the often referenced David Report was, "I became heavily involved, with many others, in helping improve the business environment of Massachusetts" (apparently at the same time he was running Sprague Electric and trying to save the world from the Russians!).

The second was the Electronics Industry Association, which was originally formed in 1924 as the Radio Manufacturers Association. He served as President and/or Chairman from 1950 to 1954 and Sprague Electric was an important and active member until the company ceased to exist in 1992. Sprague headed many of the electronic components standards and technical committees, as well as the marketing committee for many years, and Robert Sprague used the Association as a platform to argue for more open access to foreign markets such as Japan, unfortunately with little success. He was the Association's only two-time recipient of its prestigious Medal of Honor (1953 and 1978).

Directly or indirectly government service dominated much of my father's interest and life outside of

Sprague Electric. This involvement began at Annapolis and culminated in 1953 when he became an expert and consultant on nuclear deterrence and continental defense followed by his work on the Killian and Gaither Committees. Such involvement did not cease with the 1958 presentation of the Gaither Report and recommendations.

Returning to his Massachusetts Institute of Technology roots, he had become a member of the MIT Corporation in 1953 and life member starting in 1955. In 1958, he joined the first Board of Trustees of the newly formed MITRE Corporation, a not-for-profit MIT spin-off which initially directed the design and construction of the United States Air Force SAGE air defense system. He served as chairman from 1969 to 1972 and honorary trustee for life starting in 1981. Two later Sprague Electric directors, Robert Charpie and David Ragone were MITRE associates.

In 1970 he became a director of the Draper Laboratory Division of MIT, which in 1973 was spun-off as the Charles Stark Draper Laboratory, and became member emeritus in 1979. Draper's research programs have always concentrated on guidance and control systems for the United States Department of Defense and NASA.

Even in his 80s, completely removed from the electronics industry, and with his defense involvement greatly diminished, my father remained busy. He and my mother rarely traveled anymore (I never remember them travelling much together as a couple), but he had a large family and enjoyed seeing both grandchildren and great grandchildren. He could be overbearing at times. When dining with his family at places such as the Country Restaurant in Williamstown or Four Chimneys in Bennington, because he insisted on paying the bill, he had the habit of telling everyone what he felt they should eat. Although he accepted rejection gracefully, he never quite understood why many of us, especially my children, wanted no part of his favorite appetizer, prosciutto and melon.

As Sprague Electric was departing North Adams in the 1980s, increasingly the city and those who had worked at the company remembered Robert Sprague with both affection and respect. Sprague's (as it was known locally) had been a nice place to work and Robert Sprague had provided good jobs and taken care of his employees. In 1983 the Berkshire Chamber of Commerce honored him with the Francis H. Hayden Award as a resident "who has made a significant contribution to the economic, social, and/or cultural improvement of any or all the (Northern Berkshire) communities".

Throughout his life he was active in the affairs of St. John's Episcopal Church and was senior warden emeritus from 1968. From 1982 to 1986 he was founding trustee, chairman, and treasurer of the Elm Tree Foundation which he had formed with Robert McCarthy to try and stem the blight that was killing all the local elms.

*Robert C. Sprague and Nikos Psacharopoulos,
circa 1983. (Author's collection)*

By far his favorite local not-for-profit was the now iconic Williamstown Theatre Festival, where he was founding trustee, chairman of the executive committee, and early on managed the financial affairs of the Festival. Founded in 1954, he loved everything about it, his relationship with the first artistic director, Nikos Psacharopoulos, the plays, hob-knobbing with the actors, actresses, and celebrities and, toward the end of his life, when Cabaret (musical events often held after one of the plays) performers would give him a private performance in his home. They never seemed to mind because he was such an attentive and appreciative audience. Of course he had always been attentive when pretty young girls were involved.

As a fund raiser for organizations he supported he had no equal, because he refused to take "no" as an answer. After all, if he believed in something it must be worth other people's money.

Although after the 1987 death of both my brother and mother, he kept his grief mostly to himself, his personality seemed to soften. When friends and family visited, he was less prone to talk about himself, and much more interested in what they were doing. Toward the end of his life, courtesy of Penn Central, he was able to remain at home with full-time care. The group of mostly middle-aged women who provided this care all loved him. He was such a nice man, so friendly, so quick to say thank you, and at times to regale them with stories of his extraordinary life. And to him, much as he had felt about his life, every one of them was beautiful.

Appendix 1
Sprague Electric Financial Tables

Table 1: Net Sales Billed and Profit After Tax, 1941 – 1946 (From Annual Reports)

Year	NSB	PAT
1941	$4,796 M	$209 K
1942	7,373 M	208 K
1943	14,469 M	548 K
1944	20,801 M	872 K
1945	16,724 M	654 K
1946	10,767 M	720 K

Table 2: NSB and PAT, 1946 – 1953 (ARs)

Year	NSB	PAT	Employees
1945	$16.7 M	$0.65 M	
1946	10.8 M	0.72 M	2600
1947	10.5 M	0.66 M	2100
1948	12.6 M	0.83 M	2500
1949	15.3 M	1.21 M	3000
1950	28.6 M	3.35 M	4700
1951	38.3 M	2.66 M	5100
1952	43.4 M	2.86 M	5900
1953	46.8 M	2.89 M	5500

SPRAGUE ELECTRIC

Table 3: NSB and PAT, 1953 – 1958 (ARs)

Year	NSB	PAT	Employees
1953	$46.8 M	$2.89 M	5500
1954	42.4 M	3.33 M	5000
1955	44.4 M	3.00 M	6000
1956	44.7 M	2.18 M	5700
1957	46.2 M	2.22 M	5500
1958	43.2 M	1.76 M	4900

Table 4: NSB and PAT; 1958 – 1966 (ARs)

Year	NSB	PAT	Employees[1]
1958	$43.2 M	$1.76 M	4900
1959	56.4 M	3.50 M	5900
1960	64.5 M	4.09 M	6400
1961	77.3 M	6.09 M	7200
1962	87.0 M	6.43 M	8200
1963	83.3 M	4.63 M	7600
1964	85.7 M	3.60 M	8100
1965	107.1 M	4.98 M	10,200
1966	141.5 M	8.71 M	12,500

---

[1] Domestic employment through 1964 and worldwide thereafter

APPENDIX 1: SPRAGUE ELECTRIC FINANCIAL TABLES

Table 5: NSB and PAT, 1966 - 1971 (ARs)

Year	NSB	PAT	Employees[2]	Overhead Expenses[3]
1966	$141.5 M	$8.71 M	12,500	$25.68 M
1967	127.4 M	3.33 M	12,300	29.33 M
1968	132.8 M	(2.83) M	12,100	29.46 M
1969	147.1 M	1.46 M	12,300	30.13 M
1970	127.5 M	(6.88) M	10,900	29.78 M
1971	117.9 M	(8.10) M	9,700	27.12 M
1972	146.7 M	(0.30) M	10,600	26.69 M

Table 6: NSB and PAT, 1971 – 1976 (ARs)

Year	NSB	PAT	OH Expenses	Employees
1971	$117.9 M	$(8.10 M)	$27.12 M	9,700
1972	146.7 M	(0.30 M)	26.69 M	10,600
1973	197.7 M	11.63 M	32.38 M	12,600
1974	214.8 M	10.16 M	34.36 M	12,900
1975	161.9 M	(10.30 M)	34.15 M	9,500
1976	199.6 M	6.85 M	36.62 M	9,700

---

[2] Worldwide
[3] SG&A (Sales, General, and Administration) plus RD &E

SPRAGUE ELECTRIC

Table 7: Financial Performance 1981 – 1987 (PCC ARs)

	1981	1982	1983	1984	1985	1986	1987
Net Sales	$438M	405	428	571	492	449	437
Oper. Inc.	54M	40	34	40	(44)[a]	(160)[b]	7
S G & A			67M	84	88	87	78

Table 8: STI Financial Performance (STI ARs)

Year	Net Sales	Net Income	Comments
1987	$437M	$3.6M	discontinue MLCCs
1988	505	13.2	recession starts second half
1989	344	(98.2)[c]	$40M IBM shortfall
1990	315	(49.3)	sell semis. & Fil-Mag; close MLCCs
'91/'92	294	(52.5)	Lindner CEO; close or sell everything else

---

[a] Includes $33M restructuring costs (plant closings and consolidations)
[b] Includes $137M reserve for asset redeployment
[c] Includes $79.7M restructuring reserve

iv

Appendix 2
Sprague Electric Planar Process Timeline

When	Who	Where	What
1925	J. Lilienfeld	USA	CuS solid state amplifier
WW II	many	world	Ge & Si for radar detectors
1945	M. Kelly	BTL	Solid State Research Lab
12/47	J. Bardeen & W. Brattain	BTL	Ge point-contact transistor
1/48	W. Shockley	BTL	junction transistor
1948	J. Little & G. Teal	BTL	PN junction in single crystal GE
1953		Philco	Electrochemical transistor
1955	W. Shockley	Schockley Semi. Lab.	photoresist patterning & SiO_2 masking
1955	J. Atalla	RCA	planar PN diode
9/58	R. Noyce & G. Moore	Schockley Semi. Lab	led "traitorous 8" to Fairchild Semiconductor
1958		Fairchild	NPN & PNP MESA transistors
9/58	J. Kilby	T I	semiconductor integrated circuit
1959	H. Loar	BTL	Si epitaxial transistor
1959	J. Hoerni	Fairchild	Si Planar NPN transistor
4/59	K. Lehovec	Sprague	p-n junction isolation patent
1961		Fairchild	Micrologic I C family
1961 – 1963		Sprague	Silicon epitaxial planar transistor (SEPT®) and IC (UNICIRCUIT®)

APPENDIX 3: ROBERT C. SPRAGUE AND NUCLEAR DETERRENCE

"Due to its complexity I was involved in the study for nearly seven months, from September 1953 until March 1954, twelve to fourteen hours a day, seven days a week. Initially I met with Admiral Arthur Radford, Chairman of the Joint Chiefs of Staff, and the heads of the services themselves, and then key government and military officers as I traveled around the country."

The information Robert Sprague shared with David Sprague in his oral history was no longer sensitive. Some of it had already appeared in the 1987 PBS documentary, "The Nuclear Age," in which Robert Sprague was interviewed. There is also a companion book to the documentary, (*War and Peace in the Nuclear Age,* by John Newhouse, Alfred A. Knopf, N Y 1989). One of the most startling revelations concerns a November 1953 exchange Robert

Sprague had with General Curtis LeMay, during a briefing at the Strategic Air Command (SAC) headquarters in Omaha, Nebraska. They were in a briefing room with some fifty of LeMay's staff when Robert Sprague asked a question concerning potential target cities in the USSR, to which LeMay replied, "Mr. Sprague, your question concerns war plans which I will not answer nor would I answer if it were from the President of the United States instead of from you – for all I know I may have a couple of Communist spies in the room!" It appears that LeMay already had SAC aircraft flying over Russia and was prepared to make an immediate preemptive strike if it appeared the Russians were planning an attack. When Robert Sprague commented that this certainly wasn't national policy, LeMay confirmed, "No, but it is my policy!" As a conservative Republican and hard-core hawk, instead of being appalled by LeMay's position, Robert Sprague felt it actually made the United States safer. "I considered LeMay an especially capable General. His people, especially the pilots and navigators, were superbly trained, and I believe he was the most professional military man we had heading up our various services at the time."

In early January 1954, Robert Sprague felt he had all the information he needed, and he began a series of presentations of his findings and conclusions, first to Admiral Radford and senior officers of all the armed services, and then to the Senate Armed Services Sub-Committee, which prepared a top-secret report for President Eisenhower. Robert Sprague was then appointed by the President as

APPENDIX 3: ROBERT C. SPRAGUE AND NUCLEAR DETERRENCE

a consultant to the National Security Council, a position he held from May 1954 through December 1957. He also served on the steering committee of the so-called "Killian Committee," named for its Chairman, Dr. James R. Killian, President of MIT and Special Assistant for Science and Technology (or Presidential Science Advisor) to President Eisenhower from 1957 to 1959. The Killian Committee was formed in the fall of 1954 and made up of fifty-eight distinguished scientists and engineers. Although its role was to continue and expand the studies already underway on continental defense, it did so primarily from the point of view of intelligence and the security of communications in time of war. Based on a suggestion of one of its members, Dr. Edwin Land of Polaroid, one of its most important contributions was development of the U-2 spy plane that flew over Russia for four years until Gary Powers was shot down by a Russian missile. It also was responsible for formation of the SAGE Air Defense System and the MITRE Corporation, on the Board of Trustees of which Robert Sprague served for many years and chaired it for two.

The most important advisory group that was formed during this period of extraordinary tension between the United States and Russia was the Security Resource Panel, better known as the "Gaither Committee," after its initial Chairman, H. Rowan Gaither, Jr. , who was a distinguished California lawyer, who had served as Business Manager for the MIT Laboratory for Electronics during World War II (which developed military radars toward the end of the War), and was then Chair of both the Rand Corporation and Ford Foundation. It was formed in April 1957 and was the largest civilian committee of its kind ever established, including ninety-one distinguished individuals with broad and varied experience on military and economic matters, and a staff of more than twenty. As defined by Eisenhower, its mission was to analyze and make recommendations concerning the nation's position relative to Russia in active and passive defense. Robert Sprague was named Co-Director and took over as Director when Gaither was stricken with cancer in September 1957. With a ten member Steering Committee, Advisory Panel, Subcommittees, Project Managers, Executive Staff, and Administrative and Secretarial Staff, management of the entity was more than a full time responsibility.

The Security Resource Panel gave its final report to the National Security Council on 7 November 1957. There were nearly seventy attendees, including President Eisenhower, Secretary of State John Foster Dulles, Secretary of Defense Neil McElroy (replacing Charles Wilson), and Secretary of the Air Force James H. Douglas (replacing Harold Talbott). The report was sobering, controversial, and is still debated today. The report's recommendations included greatly improved protection of the United States' strategic assets, both military and civilian; expansion of its nuclear ballistic arsenal, increased capability to wage limited military operations (more than fifty years later a major priority), a reorganized Defense Department, and a major civilian fallout shelter program. The estimated cost of the program was more than $44 billion, to be spread over four to five years. Excluding fallout shelters and limited war capability, Eisenhower apparently agreed with the bulk of the report, although subsequent Presidents, including John F. Kennedy, had different reactions.

APPENDIX 4: CHRONOLOGY OF IMPORTANT DATES IN ROBERT C. SPRAGUE'S LIFE

Robert C. Sprague (3 August 1900 – 28 September 1991)

Education
    Attended Hotchkiss School, Class of 1918
    U. S. Naval Academy: graduated 1920
    U. S. Naval Post-Graduate School, B. S. 1922
    Massachusetts Institute of Technology (MIT), M. S. 1924

Sprague Specialties Company/ Sprague Electric Company
    Founder, CEO, President, Chairman 1926 -1953
    CEO, Chairman 1953 – November 1971
    Treasurer, 1954 – 1965

Honorary Chairman and Chairman Executive Committee, November 1971 – December 1976 (acquisition of Sprague Electric by General Cable/ GK Technologies)
Honorary Chairman and Director December 1976 – 1981 (acquisition of General Cable/GK Technologies by Penn Central Corporation)

Government Service

Nominated as Undersecretary of the Air Force in the Eisenhower administration 12 January 1953; withdrawn four weeks later
Study of U. S. continental defenses for Senate Armed Services Preparedness Sub-Committee, 1953 – 1954
Consultant on continental defense to National Security Council, May 1954 – December 1957
Consultant to Killian Committee, 1954 – 1955
Director Gaither Committee, 1957 – 1958

Universities, Organizations, Associations

MIT, Member of Corporation from 1953 (life member from 1955)
Mitre Corporation, Founding Trustee, from 1958 (Chairman, 1969 – 1972)
Charles Stark Draper Laboratory, Director from 1970 – 1978; member emeritus from 1979
Associated Industries of Massachusetts, from 1945, President 1951 - 1953

Massachusetts Science and Technology Foundation, Governor 1970 – 1978

First National Bank of Boston, Director 1961 – 1973

Northeastern University, Member of Corporation from 1953

Federal Reserve Bank of Boston, Chairman and Director, 1955 – 1960

Electronic Industries Association (EIA), Governor from 1943 (various terms as President and Chairman)

IEEE, from 1941 (Fellow from 1949, Life from 1971)

American Academy of Arts and Sciences, Fellow from 1960

Hudson Institute, Founding Trustee 1961 – 1971

Williamstown Pine Cobble School, Trustee from 1942 (Chairman 1942 – 1968)

St. John's Episcopal Church (Senior Warden Emeritus from 1968)

Elm Tree Foundation (1982 – 1986), (founding Trustee, President and Treasurer)

Williamstown Theatre Festival, from 1954 (Founding Trustee and Chairman of Executive Committee)

Honorary Degrees

Northeastern University, Doctor of Engineering, 1953

Williams College, Doctor of Science, 1954

Tufts University, Doctor of Laws, 1959
Lowell Technology Institute, D S, 1959
University of New Hampshire, D L, 1967
North Adams State College, D S, 1972
University of Massachusetts, D S, 1975

Other Honors and Awards

EIA Medal of Honor, 1954 and 1978 (only two time recipient)

Hotchkiss Alumni Association "Man of the Year" Award, 1958

New England Council "Man of the Year" Award, 1965

Francis Hayden Award (North Adams, MA), 1983

American Academy of Engineering, elected Member 1985

ENDNOTES

[1] For more on Frank J. Sprague see *Engineering Invention, Frank J. Sprague and the US Electrical Industry* by Frederick Dalzell (MIT Press, Cambridge, MA 2010), *Frank Julian Sprague, Electrical Inventor & Engineer*, by W. D. Middleton and W. D. Middleton, III (Indiana University Press, Bloomington, IN 2009), and *The Birth of Electric Traction* by Frank Rowsome, Jr (IEEE History Center Press, 2013)

[2] "David Report", 1989 (JLS Personal Papers)

[3] JLS personal papers

[4] For more detail on how a capacitor is manufactured and used, see *Revitalizing US Electronics, Lessons from Japan* by John L. Sprague Butterworth-Heinemann, Stoneham, MA 1993)

[5] An electrolytic capacitor has one conducting plate or anode, usually very high purity aluminum or tantalum, on which a thin metal oxide is formed as the dielectric and a conducting electrolyte (which can be a liquid, "dry", or "solid") for electrical connection

to the cathode, which is often a metal can. The "dry" aluminum electrolytic was licensed in the early 1930s from P. R. Mallory, a Sprague competitor, under its Sam Ruben patents.

[6] *North Adams Transcript*, 9 August 1948

[7] *A Study of Union History at the Sprague Electric Company in North Adams, MA, 1929-1970*, by Raymond C. Bliss '76, pp 9-14 (Unpublished Thesis in partial fulfillment of requirements for BA in History from Williams College)

[8] *North Adams Transcript*, 22 September 1943

[9] Ibid, 24 September 1945, 1 October 1945

[10] Ibid, 25 November 1944

[11] Ibid, 31 August 1944

[12] Bliss, p. 33

[13] David Sprague interview of his grandfather (endnote 2)

[14] March 9, 1946 Sprague Log

[15] For more on the history of the transistor see pages 18 and 19 of *Revitalizing* (endnote 4) and *History of Semiconductor Engineering* by Bo Logic (Springer-Verlag Berlin Heidelberg 2007.

[16] Bo Lojek (endnote 15); Kurt's colorful remembrance of the hearing can be found on pages 202 – 205.

[17] In epitaxy a thin lightly doped single crystal layer of, for example, Si is grown on the surface of a heavily doped Si substrate. This causes reduction of the resistance of the collector region of a MESA transistor, thus eliminating the performance advantage held by the electrochemical transistor.

[18] *Fundamentals of Semiconductor Devices,* by Joseph Lindmayer and C. Y. Wrigley (D. Van Nostrand and Co., 1964)

[19] 1960 Annual Report

[20] 1968 Annual Report

[21] For a while the semiconductor division had its own sales organization. However, this was eventually given up as too expensive and because it created too much internal friction within the company.

[22] *We Came in Peace For All Mankind*, by Tahir Rahman (Leather's Publishing, 2008)

[23] Mitre is a large private corporation founded in 1958 to provide direction of the United States Air Force SAGE air defense system. Heavily defense oriented, over the years it has managed a variety of Department of Defense early warning and communications systems

[24] Berkshire Eagle, 6/18/87; the third largest was the North Adams Hospital with 500 employees

[25] *Passive Component Market Outlook, 2008 – 2013,* Paumonok Publications, Inc., Cary, NC 2008

[26] Berkshire Eagle, 8/11/99

INDEX

Page references in italics indicate photographs. Numbers followed by n indicate endnotes.

A

acquisitions: aborted, 101, 104–106, 154; by General Cable, 101–102, 107–109, 154; of Herlec Corporation, 48; by Penn Central, 101–102, 113–114, 117, 154; by Sanken Electric, 133–134; United Technologies purchase of Mostek, 112–113; by Vishay, 135–136

Actions for Profitable Growth, 120–121, 125–126

AFL-CIO, 26, 78–79

AFTE Local #101, 78–79, 89, 91

Allegro Microsystems, LLC, 98, 134, 136–138

aluminum capacitors, 29–30, 60, 135

Aluminum Electrolytic Capacitor, 18

aluminum electrolytics, 16, 25, 178n5

Aluminum Electrolytics, 42

Alvarez-Tostado, Claudio, 65–66

American Academy of Arts and Sciences, 175
American Academy of Engineering, 176
American Premier Underwriters, Inc., 135
Apollo 11, viii, 87–88
Armitage, Bob, 97
arms race, 58–61
Army-Navy "E" Award, vii, 30, 33–34, *34,* 35
Arnold Print Works, ix, 2, 36, 139, *140*
Arnold Town, ix
Arvida Corporation, 114
Associated Industries of Massachusetts, 43, 155, 174
Atom capacitors, 27
atomic bomb: deterrence, 57–58, 167–171; trigger mechanisms, vii, 35–36
AT&T, viii, 50
Automation Industries, 109
automotive components, 109, 137
Autonetics (North American Aviation), vii–viii, 76–77
Avery, Mary E. ("Molly"), 10–11
AVX, 77, 103, 133, 136, 139

B
Bank of New York and Trust Company, 6
Bardeen, John, 50
barium titanate (BaTiO3), 61
Barker Microfarads, 135
Barre, Vermont, 42, 47, 79
Barrett, John, 90, 121, 141
Bateman, George, 88
batteries, 48

INDEX

Beacon Realty and Trading Company, 36
Beaver Mills, 16–17, 19
Beaver Street plant, ix, 20, 29, 42, 73
Bell Telephone Laboratories, viii, 50, 53, 60
Berkshire Capital Investors (BCI), 143
Berkshire Chamber of Commerce, 157, 176
Berkshire Daily, 143
Berkshire Eagle, 95–96
Bernard, Walter J., 68–69
Berner, Warren, 72
Bevis, John, 12
Bliss, Raymond, 29, 38–39
bombs: atomic, vii, 35–36; incendiary, 30, 32; nuclear deterrence, 57–58, 167–171
Bond, Frank, 19, 63
Boxcar Media, 143–144
Brafman, Harold, 12
Brattain, Walter H., 50
Briggs, Taylor, 143
Browne, Tom, 123, 125
Brown Street plant, ix, 12, 24, 28–33, 37, 42, 73, 122
Burroughs, 21
Busen, Carl, 67–68

C
Cabot Corporation, 100–101, 104–106, 154
calculators, handheld, 85–86
California operations, 42
    capacitors, x, 9, 12–13, 35–36, 47–49, 63, 115, 177n5; aluminum, 29–30, 60, 135; assembly, *43*; Atom, 27;

ceramic, 48, 60–61, 63, 77, 103–104, 133–134, 138–139; coupling, 47–48; decoupling, 48; demand for, 40; film capacitor companies, 135; HYREL®, 77; Midget®, vi, 11–13; MONOLYTHIC®, 61; multi-layer, 60–61, 63, 77, 103–104, 108–109, 133–134, 138; parallel plate, 13; TANTALEX®, 59–60; tantalum, 59–60, 77, 126, 128, 136, 139

Carey, Amos, 29

Carlson, Bruce, 94, 97, 100, 102–103, 105

Carrolton, Texas, 85

Casey, James, 68

CERACIRCUITS® (thin film ceramic-based hybrid circuits), 70

ceramic capacitors, 48, 60, 138–139; multi-layer, 61, 63, 77, 103–104, 133–134, 138

Ceramic Division, 74

Charlestown Naval Shipyard, 8

Charpie, Robert, 99–100, 104–105, 107–108, 156

Christiansen, Don, 124

cigars, 110

Clark Art Institute, 57

Cold War, vii–viii, 40–41, 57–61

Collegiate School, 3

Columbia Radio Corp., 21

Commonwealth Sprague, 122

competition, 126, 146. *see also* Japan

Concord Semiconductor Plant, 42, 54, 81, 94, 103, 128, 136

condensers: fixed-paper, 9–10; "Midget," 11

Connolly, Arthur G. (Art), 94–95, 97, 106–108

consumer sales, 14
continuing education, 71
Cornell-Dubilier, 11, 24–25
corporate strategy, 27–28, 98, 126; niche strategy, 80, 109, 138, 152; second-source strategy, 75–76, 79–80
Counts, J. Curtis, 90
coupling capacitors, 47–48
CPower, 143
Crane & Co., 144
Cushman, Norton, 70

D
Dallas Semiconductor Plant, 113
Darcy, Jack, 123–124
Dearborn Electronics, 135
decoupling capacitors, 48
defense interests, 30–31, 58
deindustrialization, 139
Delco Radio, vii–viii, 25, 83, 96
Dicker, Richard, 115
Digital Equipment, 113
Diversified Industries Group, 114
doldrums, 58–61
Doolittle, James, vii, 32
Douglas, James H., 170–171
DRAM (dynamic random access memory), 112
Charles Stark Draper Laboratory, 156
Dreiner, Rudolph, 68
Duca, Bob, 85
Dukakis, Michael, 9, 141

Dulles, John Foster, 170–171
DuPont, 61
Duracap®, 47
dynamic random access memory (DRAM), 112

E
"E" Awards, vii, 30, 33–34, *34,* 35
Eisenhower, Dwight D., 55–56, 168–171
electrolytic capacitors. *see* capacitors
electron-beam addressable memories, 98–99
Electronic Industries Association (EIA), 62, 155, 175; Medal of Honor, 113, 155, 176
electronics industry, v, 22, 40–41, 63, 91
Elm Tree Foundation, 132, 157, 175
Emergon, 143
Emerson Electric, 25
enamel, vitreous, 61
entertainment electronics, 22, 63
E R Corporation, 98–99
Erhard Seminars Training, 125
Erickson, Joseph A., 63, 93, 97
etch house, 20–21
Evans, J. S., 33
Everyday Health, 143
Experimental Laboratory, 20
Eziba, 143

F
Fabmica®, 12
Fabricius, Jack, 61

Fairchild Camera and Instrument, 54
Fairchild Semiconductor Corporation, 54, 66, 75–76
Farmer, A. A., 33
Federal Reserve Bank of Boston, 175
Fil-Mag, 133–134
film capacitor companies, 135
Filter Division, 64, 122, 133–134
First National Bank of Boston, 175
Fitzgerald, Dennis, 137
fixed-paper condenser, 9–10
fixed resistor products, 27
Flood, George B., 19, 23
FM radio, 28
Ford Motor Company, 25
Fortune 500, 77
Fourth Decade Committee, 62
Fowkes, Fred, 70

G
Gaither, H. Rowan, Jr., 169–170
Gaither Committee (Security Resource Panel), 156, 169–171
Gaither Report, 156
Galvin, Bob, 76
gas masks, 28, 30
Geekcorps, 143
GE Labs, 66
General Cable Corporation, 101, 106–114, 138, 154
General Electric (GE), 21, 26–27, 66
General Motors, viii, 83

GE-Pittsfield, 122
germanium, 151
Germany, 129–130
GK Technologies, 109, 113–114
government service, 45, 54–58, 78, 147, 155–156, 167–171, 174
Great Depression, 24–25
Great Southwest Corporation, 114

H

Hall Cells, viii, 109
Hall-effect magnetic sensors, 137
handheld calculators, 85–86
Hanschen, Richard, 84–85
Harriman-and-West Airport (North Adams, Massachusetts), 119
Harris, Matt, 143
Francis H. Hayden Award, 157, 176
Healthshare, 143
hearing aid batteries, 48
Herlec Corporation, 48
Hershey, Brett, 142–143
high-fidelity audio speakers, 98–99
Hogan, Les, 76
home entertainment systems, 22, 63
Hong Kong, 92
Hoosac Worsted Mill, 24
Hotchkiss Alumni Association, 176
Hotchkiss School, 3–5

House Subcommittee on the Impact of Imports and Exports on American Employment, 78
Hudson Institute, 175
Hughes Aircraft Thin Film Laboratory, 66
Hurley, William R., 16
Husher, John D., 76, 81
Hutchinson, Eric, 68
Hyatt, Antoinette Gray ("Nettie"), 6
Hybrid Systems, 97–98
HYREL® capacitor, 77

I
IBM, vii–viii, 21, 48, 60, 96, 126–128
IEEE, 175
incendiary bombs, 30, 32
Independent Condenser Workers (ICW): ICW #1, 26–27; ICW #2, 26–27, 37–39
inductors, 49
integrated circuits, 51–53, 68, 109; linear, 94, 109; silicon-based (UNICIRCUITS®), 68, 70
International Association of Machinists, 38–39
International Union of Electrical Workers (IUE), 78, 88–92
inventions, 9, 22
ion implantation, 68
iPowerPlay, 144
Ishler, Ken, 76
Israel, 135–136
Italy, 42

J

Japan, 77–78, 86, 98, 103–104, 112, 126, 155
Jensen, Robert P. (Bob), 106–110, 113, 116, 154
John L. Sprague, Associates, 131
Jones, Harriet Chapman, 44
Joseph A., 107–108
Juarez, Mexico, 94

K

Kalker, Harry, 14
Kampman, Donald R., 107–108
Kelley, Bob, 88
Kelley, Mervin J., 50
Kemet, 60, 133, 136, 139
Kemex, 103
Kennedy, George, 143
Kennedy, John F., 171
Keyes, Robert, 55
Kilby, Jack, 52
Killen, Carroll, 64
Killian, James R., 58, 169
Killian Committee, 156, 169
Kimball, Allan, 137
Kinzel, Gus, 97, 107–108
Korean War, vii–viii, 40
Kosnik, Ed, 125, *127,* 128, 130–134
Krens, Thomas, 140–141
Kyocera/AVX, 139

L

labor rates, 27, 89–90, 92, 135–136
  labor relations, 26–27, 38–39, 78–79, 88–100, 138, 153
labor unions, 26–27, 38–39, 78–79
Lack, Frederick R. (Fred), 62, 97, 99
Lalley, H. Fred, 16
Land, Edwin, 169
Lazier, Wilbur, 18
Lee, Len, 98
Lehovec, Kurt, viii, 51–53, 62, 66, 72, 149
LeMay, Curtis, 168
Lesher, Bob, 5
Lexington, Massachusetts, 120–121, 124
Leyden Jars, 12
Lindmayer, Joseph, 68
Lindner, Carl H., Jr., 102, 115–116, 118, *127,* 128–129, 131, 135, 139
linear integrated circuits, 94, 109
liquid crystal displays, 98–99
Little, Dennis G., 107–108
Living and Leisure Group, 114
Loconto, Peter, 117, 123
The Log, 21, 27, 29–32, 57, 150
Lowell Technology Institute, 176
Lycos, 143

M

M514 proximity fuze program, 75–76

MacDougall, John, 68, 85
MacVicar, Margaret, 99–100, 102–103, 107–108
Maden, Peter, 124–125
Magnavox, 21
Mahar, Galeb, 72
Mahar, Hal, 124
mahogany row, 72
Malaysia, 92
Manchester, Kenneth (Ken), 68, 81, 85
Manhattan Project, 35–36
Manhattan Square, 35–36
Marathon Manufacturing, 114
Marsh, Howie, 72
Marshall Street complex, ix, 29–30, 42, 73, 102, 120, *140*; Building 1, 72; Building 4, 72; Building 6, 74; conversion to MASS MoCA, 140–142; mahogany row, 72; temporary closure, 37
Martinelli, Al, 115, 120, 123, *127*
Massachusetts College of Liberal Arts (MCLA), 71–72, 144
Massachusetts Institute of Technology (MIT), 7–8, 71, 156, 169–170
Massachusetts Miracle, 141
Massachusetts Museum of Contemporary Art (MASS MoCA), x, 22, 72, 139, *140,* 140–142, 144
Massachusetts Science and Technology Foundation, 175
MASS MoCA Foundation, 141
McCarthy, Robert, 157
McElroy, Neil, 170–171

McGill, Robert, 143
McGuiness, Don, 123–125
McKinsey, 129
McLean, William E. (Bill), 80, 106, 108
mega death, 57
mergers, 133. *see also* acquisitions
MESA-junction transistors, 151
mesa technology, 53–54
Metal Oxide Semiconductor (MOS) devices, 84–85
Mexico, 81
mica, 12–13, 60–61
Micro-Bit, 98–99
Midget® capacitor, vi, 11–13
Milan, Italy, 42
military sales, 59
Milton, Bill, 124
Minuteman ICBM guidance and control system, vii–viii, 76–77
missiles, vii–viii, 76–77
MIT (Massachusetts Institute of Technology), 7–8, 71, 156, 169–170
MIT Corporation, 156, 174
MITRE Corporation, 58, 99, 156, 169, 179n23
MONOLYTHIC® capacitors, 61
moon wafers, 86, 86–88
Morris, Larry G., 107–108
Morrison, Dick, 131, 137
Morton, Thomas, 95
MOS (Metal Oxide Semiconductor) devices, 84–85
Mostek, viii, 84–86, 93, 108–109, 112–113

Mostek Model, 98, 153–154
Motorola, 76
MRA Labs, 135
multi-layer capacitors, 60–61, 63, 77, 103–104, 108–109, 133–134, 138
Multiple Semiconductor Assembly, 51
Murata, 77
Murphy, John, 123

N
NASA, 87, 156
Nashua, New Hampshire, 42
National Academy of Engineering, 132
National Security Council, 168–171
NEC/TOKIN, 136
Nejame, Mitchell F., 28
New Business Resources, 84–85
Newcomen Society, 29
New England Council, 176
Newhouse, John, 167
"new Sprague," 64–65, 112
New York Central Railroad, 114
New York Life Insurance and Trust Company, 6
New York Stock Exchange, 77, 107
niche strategy, 80, 109, 138, 152
Nobel Prize, 50, 52–53
Nolan, William J. (Bill), 14, 16, 97
North Adams, Massachusetts, v–x, 2, 19, 47, 67, 120–122; deindustrialization, 139; future directions, 144; Harriman-and-West Airport, 119; population, 139

INDEX

North Adams Industrial Company, 16
North Adams Research Center, 70, 94
North Adams Research Lab, 85
North Adams State College, 71–72, 176
North Adams Transcript, 19, 29–30, 33, 38, 107
North American Aviation, vii–viii, 76–77
Northeastern University, 175
"The Nuclear Age" (PBS), 167
nuclear deterrence, 57–58, 167–171

O

"old Sprague," 64, 112
Olsen, George, 61
Olsen, Ken, 113
OPEC oil embargo, 98

P

P. R. Mallory Co., 24–25, 178n5
Palmer, Bob, 113
parallel plate capacitors, 13
Parallel Skiing for Weekend Skiers (Sprague), 146
patents, 18, 51–53, 60, 62
PBS, 167
Peabody, Bo, 142–143
Penn Central Corporation (PCC), 101, 113–132, *127,* 135, 139, 154, 159
Penn State, 71
Pennsylvania Central Railroad, 114
Pepper, Robert S. (Bob), 70–71, 81, 83, 87–88, 152
Pepper, Star, 71

Petritz, Richard (Dick), 84–85
Phelps, Gordon, 97
Philco Corporation, 21–22, 25–26, 53, 151
Philippines, 92
planar technology, 53–54, 152, 165–166
planar transistors, 67–68, 94
PN Junction Isolation, viii, 62
Polaroid SX-70 camera, 83
Ponce, Puerto Rico, 42
Powers, Gary, 169
PRG, 143
pricing, 64
Princeton Material Sciences, 98–99
product life cycles, 24
Prothro, Vin, 113
Provost, A. Normand (Norm), 81
proximity fuzes: M514, 75–76; VT, vii, 34–35, 48
Psacharopoulos, Nikos, 158, *158*
Puerto Rico, 42, 92

Q
Quincy, Massachusetts, 2, 8–9

R
Radford, Arthur, 167–168
radio, vi, 28
Radio Manufacturers Association, 155
Ragone, David, 99–100, 105, 107–108, 156
random-access memory, 85–86
Raytheon, 21

RCA, 26
Renaix, Belgium, 82
research and development, 24–25, 69–74
Resistor Division, 64
resistors, 49
Resounding Technology, 143
retail sales, 14
revenues, 126–127
Rhedyt, Germany, 82
Rickover, Hyman, 17
Robbins, Harry, 63
Robinson, Preston, 18, 35–36, 51, 60, 62
The Rock of Ages Capacitor Corporation, 33
Rosen, Ben, 113
RPI, 71
Ruben, Sam, 178n5
Russia, 57–58, 155, 168–169

S

S-2000 Transac (Philco), 53
Sabor, Richard, 142–143
SAGE Air Defense System, 156, 169
Saltonstall, Leverett, 33, 57–58
Sanken Electric Co., Ltd., 134, 136–137
SBE, Inc., 135
Scarborough, Fred, 64
Schaffhausen Museum, 140
Scheer, Hans, 68
Scotland, 82
Seacord, John, 81

Searls, David C., 107–108
Security Resource Panel (Gaither Committee), 156, 169–171
Semiconductor Division, 64, 85, 179n21
semiconductors, viii, 52, 62, 65–71, 112, 132–134, 138, 148–153; Concord Semiconductor Plant, 42, 54, 81, 94, 103, 128, 136; Dallas Semiconductor Plant, 113; Fairchild Semiconductor Corporation, 54, 66, 75–76; Metal Oxide Semiconductor (MOS) devices, 84–85; Multiple Semiconductor Assembly, 51; silicon wafers, 94, 116; Sprague Semiconductor Group, 80, 83–84, 98, 103–104, 125, 136–137; "Studies on the Nature of Metal to Semiconductor Alloy Junctions" (Sprague), 66; Worcester Semiconductor Plant, 74–88, 75, 93–98, 102–103, 109, 116–117, 128, 152
SEPT® (silicon epitaxial planar transistor), 68, 94
Sevin, L. J., 84–85, 113
Sevin Rosen Funds, 113
Sharif, Lou, 84–85
Sharon, Connecticut, 5
Sherry, Jim, 124
Shockley, William B., 50
Shockley Labs, 66
Shugg, Carleton (Carl), 17, 27, 29
Signetics, 75
silicon-based integrated circuits (UNICIRCUITS®), 68, 70
silicon epitaxial planar transistor (SEPT®), 68, 94
Silicon Village, 142–143

silicon wafers, 94, 116; moon wafers, *86,* 86–88; Worcester Wafer Fab, *82*
Six Flags Corporation, 114
Skillview, 143
Smith, Doug, 123
Solid State Research Laboratory, 50
solid-state revolution, 49–100
Solid State Scientific, 117
Solid State Scientific, Aluminums, 128
solid tantalum capacitors, 77, 126–127, 133–136
Solid Tantalums, 125, 133–134
South East Asia, 81
Space Age, vii–viii
Space Shuttle, viii
Special Products Division, 64
spin-offs, 101
Sprague, Althea, 15, *15*
Sprague, Bill, 66–67
Sprague, David, 3, 43, 167
Sprague, Florence A., *8,* 9–11, 131, 153
Sprague, Frank Desmond, 2, 14–15, *15,* 43
Sprague, Frank J., vi, 2, 5, 14, *15,* 18
Sprague, Harriet, *15*
Sprague, Helene, *15*
Sprague, Jid, 146, 148
Sprague, John L., Jr., 66–67, *86, 111,* 148
Sprague, John Louis, 45–46, 52, 56, 64–67, 72, 82–83, *97,* 102–104, 106; birth, 8, 65; as CEO, 113, 121, 123–126, 154; as COO, 154; and Darcy, 124; education, 146–149; and General Cable, 108; and

Jensen, 110; and Kosnik, 129–130; labor relations, 90; and Penn Central, 116–118; as President, 129–131, 154; research activities, 65–66, 68–69; and semiconductors, 65–66, 79–84, 93–95; severance, 131; and Robert Sprague, 83–84, 145–159; "Studies on the Nature of Metal to Semiconductor Alloy Junctions," 66; in US Navy, 147; and Welch, 96, 112

Sprague, Julian King, 2, 5, 11, *15,* 16, 43–45, 54, 62, 101; death, 23, 45; inventions, 22; and Robert Sprague, 57

Sprague, Robert Chapman ("R. C."), vi, 2–9, *8, 15,* 43–44, 61–62, *69, 97,* 101, *158;* associations, 174–175; business career, 17–18; as CEO, President, Chairman, 14, 23, 55, 105–106, 149–155, 173; and Charpie, 105; as conservative Republican, 168; as consultant to National Security Council, 168–169; corporate strategy, 75–76, 98; death, 132; "E" Award, vii, 30, 33–34, *34,* 35; education, 7–8, 173; as elder statesman, 155–159; as father, 145–149; and Frank Sprague, 5; government service, 45, 54–58, 78, 147, 155–156, 167–171, 174; Francis H. Hayden Award, 157, 176; as Honorary Chairman, 95, 153–154, 174; honorary degrees, 175–176; honors and awards, 176; important life dates, 173–176; inventions, 9, 22; John Sprague's reflections on, 145–159; and Julian, 57; labor relations, 88, 90; Medal of Honor, 113, 155, 176; nomination as Undersecretary of the Air Force, 45, 54–57; nuclear deterrence, 57–58, 167–171;

organizations, 174–175; *Parallel Skiing for Weekend Skiers,* 146; resignation from Sprague, 95–96; salary, 16; and semiconductors, 75, 82–84; stock interest, 55–57; testimony to House Subcommittee on the Impact of Imports and Exports on American Employment, 78; as Treasurer, 14, 173; as trustee, MITRE Corporation, 58, 169, 174; universities, 174–175; in U.S. Navy, 3–4, *4,* 5–8, 146

Sprague, Robert Chapman ("R. C."), Jr. (Bob, Jr.), 45–46, 54–55, 62, 72, 95, 97; birth, 8, 10; death, 46, 131, 153; labor relations, 88, 90

Sprague, Ruth, *15*

John L. Sprague, Associates, 131

Sprague Electric Clock Tower, *111*

Sprague Electric Company, 41–49, 101–144; acquisitions, 48, 101–102, 107–109, 113–114, 117, 133–136, 154; Annual Reports, 39, 91, 104–105; Board of Directors, 16, 61–63, *97,* 99–100, 102–103, 153–154; break-out, 61–69; capacitors, *43,* 47–48, 77; Commonwealth Sprague, 122; consulting relationships, 71; contributions to war effort, 28–32, 34–37, 39; corporate strategy, 27–28, 75–76, 79–80, 109, 138, 152; downsizing, 129–130; employee shares, 26–27; employment, 14, 37, 41, 63, 71, 119, 121–122, 129; Executive Committee, 93, 95, 105–106; expenses, 22–24, 42, 59, 79, 125–126; financial performance, 79–80, 98–99, 108–109, 118–119, 125, 128, 161–164; as flagship of Penn Central, 123; as Fortune 500 company, 77; Fourth Decade Committee, 62; growth and expansion, v, vi–vii, 19, 36, 40, 44, 118–121; headquarters, 121–122; history, v–vi, vi–viii, 2; income or revenues, x, 16, 41, 63, 79, 93, 96, 108–109; investments, 98–99; labor relations, 26–27, 36–39, 78–79, 91, 102–103; layoffs, 93; location, 14, 18–19; M514 proximity fuze program, 75–76; management, 22–23, 64–65; and Mostek, viii, 84–86; Multiple Semiconductor Assembly, 51; "A New Beginning," 130–131; "new Sprague," 64–65; niche strategy, 80, 109, 138, 152; NYSE listing, 77; "old Sprague," 64; operating divisions, 63–64; patents, 18, 51–53; planar technology, 53–54, 165–166; planar transistors, 67–68, 95; plants and facilities, ix–x, 42, 47–48, 59, 77, 82, 119; pricing, 64; products, 24, 27–30, 42, 49; profits, 23–25, 28, 39, 41, 63, 82, 96, 99, 125; as publically owned company, 99; reconversion, 36–37; relocation, 19, 121–123; research and development, 24–25, 59, 69–74; salaries, 91; sales, 14, 16, 20, 22–25, 28–30, 39, 59, 82, 91, 99, 108–109; Semiconductor Research Department, 72; semiconductors, 51–54, 59, 62, 70–71, 86; SEPT® (silicon epitaxial planar transistor), 67–68, 95; solid-state revolution, 49–100; Solid Tantalums, 125; stock interests, 55; structure, 105–106, 130; subcontractors, 33, 37; successor businesses, 135; tantalum capacitor market share, 60; transistors, 49–54, 64; UNICIRCUITS®, 68; VT proximity fuze project, vii, 34–35, 48; wages and benefits, 27, 89–90, 92. *see also* Sprague Specialties Company; *specific divisions and plants*

Sprague Flight Operations, 46
Sprague house (Quincy, Massachusetts), *1*
Sprague Products, 14
Sprague Products Distributor, 108–109
Sprague Semiconductor Group, 80, 83–84, 98, 103–104, 125, 136–137
Sprague Specialties Company, vi, ix, 2, 9–31; contributions to war effort, 28–32, 34–37, 39; employees, 10–11, 19–20, 28, 30, 33; expenses, 22–24; income or revenues, 16; location, 14, 18–19; name change, 36; net worth, 16; products, product lines, and product families, 24, 27–30; profits, 23–25, 28; reputation, 25; research and development, 24–25; sales, 14, 16, 20, 22–25, 28–30; Sales Office, 20; subcontractors, 33
Sprague Technologies, Inc., 101, 123, 132–139; Annual Reports, 132–134; Board of Directors, 128–129, 131, 135; Components, 132–133; Fil-Mag (Filter Division), 133–134; financial performance, 132–134; headquarters, 123; Semiconductors, 132–134; Solid Tantalums, 133–134
Sprague Town, ix
Sputnik, 57
St. John's Episcopal Church, 157, 175
Stackpole, William, 37
Stanford University, 71
start-ups, 143
Steinberg, Gerry, 37
Stevenson, Jordan and Harrison, 22–23
STI. *see* Sprague Technologies, Inc.

Storey, 144
Strategic Air Command (SAC), 168
Streetmail, 143–144
strikes, 26–27, 38–39, 79, 88–100, 138, 153
"Studies on the Nature of Metal to Semiconductor Alloy Junctions" (Sprague), 66
subcontractors, 33, 37
surface-barrier transistors, 53
surface-mount technology, 133–136
Swartz, Gene, 116, 120–121
Switzer, Larry, 123–124
synthetic dielectrics, 61

T

Taiwan, 82, 92
Talbott, Harold, 55–56, 170–171
TANTALEX® capacitors, 59–60
tantalum capacitors, 60, 126–128, 133–136, 139; HYREL®, 77; Solid Tantalums, 125, 133–134; TANTALEX®, 59–60; Wet and Foil Tantalums, 122
Taylor, Dr., 67
television, 28, 40–41, 83
Texas Instruments, 52, 75, 84–85
thin film ceramic-based hybrid circuits (CERACIRCUITS®), 70
Thompson, Joseph C., 141–142
Thomson Semiconductors, 113
Tiger International, 116
titania (TiO_2), 61

Tone Control, 9–10, *10,* 11
Tours, France, 82, 136
Transistor Division, 64
transistors, 49–54; MESA-junction, 151; planar, 67–68, 94; SEPT®, 68, 94; surface-barrier, 53
"Transistor Symposium" (Western Electric), 51
Tremblay, Gerry, 108
Tripod, 142–143
Tucker, Althea, 15, *15*
Tucker, Bud, 15, *15*
Tufts University, 176
Twin Lakes, Connecticut, 5
Type 31 C MONOLYTHIC® capacitor, 61

U

U-2 spy plane, 169
UNICIRCUITS® (silicon-based integrated circuits), 68, 70
Union Street plant, 73
United Chemi-Con, 135
United Electrical, Radio, and Machine Workers of America (UE), 26–27, 38–39; Local #249, 36–37
United States: Department of Defense, 156, 170; Department of Navy, vii; electronics industry, v, 63, 90–91; manufacturing sector, 102; Senate Armed Services Sub-Committee, 168; space program, 87
United States Air Force, 156, 169
United States Army, vii
United States Naval Academy, 3–4
United States Naval Post-Graduate School, 7–8

United States Navy, vii; "E" Award, vii, 30, 33–34, *34,* 35; VT proximity fuze, vii, 34–35, 48
United States Patent Office, 52–53
United States Signal Corps, 51
United Technologies, 112–113
University of Connecticut, 71
University of Massachusetts, 176
University of New Hampshire, 71–72, 176
USS Helena, 35
USS Hornet, 32
USS Kleinsmith, 56, 65, 147
USS Lexington, 8
USS Northampton, 65
USS Pennsylvania, 7

V

van Zelm, Florence Antoinette, 5–7
van Zelm, Hugh H., 107–108, 147
van Zelm, Johannes Louis, 6–7
VeFAC Analysts, 80
VeFAC Program, 80, 90
Vermont, 42
Vietnam War, 63
Village Founders, 143
Village Ventures, 143
violence, 89; 1970 strike, 90
Visalia, California, 42, 79
Vishay Intertechnology, Inc., 135–136, 139
Vishay-Sprague capacitors, 135–136
Visivox, 22

vitreous enamel, 61
Vogel, F. Lincoln ("Linc"), 70
VoodooVox, 143
VT proximity fuzes, vii, 34–35, 48

W
Walker, Ron, 116
walkouts, 38
Wall, James E., 28
Wall-Streeter Shoe Company, 28
War and Peace in the Nuclear Age (Newhouse), 167
Ward, Ernest L. (Ernie), 46–47, 97, 105
Welch, Neal, 23, 64, 72–73, 80, 96, 97, 100–106, 111, 153; as Chairman and CEO, 154; and General Cable, 108; and Jensen, 110; and John Sprague, 112
Western Electric, 50–51, 60, 96
West Germany, 129–130
Westinghouse, 21, 25–26, 76
Wet and Foil Tantalums, 122
Whitney, Jid, 147
Wichita Falls, 103
Wied, Otto, 72
Wiesner, Jerome B. (Jerry), 63, 97, 99
Wilhelmina, 6–7
Williams College, 72, 122, 142, 144, 175
Williams College Museum of Art, 140
Williamstown Pine Cobble School, 175
Williamstown Theatre Festival, 132, 158, 175
Wilson, Charles E., 55–56, 170–171

Winant, John, 88
Windover, Fred, 123
Winokur, Herbert S. ("Pug"), Jr., 116, *127*
WOFAC Corporation, 80
women directors, 99–100
women employees, 19–20
Wood, Walter, 88
Worcester Semiconductor Plant, 74–88, *75,* 93–96, 102–103, 152; chip-and-wire hybrid circuit operation, 97–98; expansion, 117; niche strategy, 80, 109; profitability, 128; silicon wafer processing, 94, 116
Worcester Wafer Fab, *82*
Work-Factor Corporation, 80
World War II, vii, 2, 27–28, 31–40, 169–170; Battle of the Bulge, 35; Sprague contributions to war effort, 28–32, 34–37, 39

X
Xtend Energy, 143

Z
Zacharias, Jerrold R., 63, *97,* 107–108
Zandeman, Felix, 135–136
Zenith, 83
Zuleeg, Rainer, 66

www.ingramcontent.com/pod-product-compliance
Lightning Source LLC
Chambersburg PA
CBHW051641170526
45167CB00001B/289